SCC Library
DISCARD
3 3065 00356 1085

D0031854

The Formation of
The Solar System
Theories Old and New

The Formation of The Solar System

Theories Old and New

Michael Woolfson
University of York, UK

Imperial College Press

Published by

Imperial College Press
57 Shelton Street
Covent Garden
London WC2H 9HE

Distributed by

World Scientific Publishing Co. Pte. Ltd.
5 Toh Tuck Link, Singapore 596224
USA office: 27 Warren Street, Suite 401-402, Hackensack, NJ 07601
UK office: 57 Shelton Street, Covent Garden, London WC2H 9HE

British Library Cataloguing-in-Publication Data
A catalogue record for this book is available from the British Library.

THE FORMATION OF THE SOLAR SYSTEM
Theories Old and New

ISBN-13 978-1-86094-824-4
ISBN-10 1-86094-824-3

Printed by FuIsland Offset Printing (S) Pte Ltd, Singapore

Contents

Introduction

Most scientists think that the work they do is very important. Well, they would wouldn't they? It is a human trait, an aspect of vanity, to consider that what one does is more significant than it really is. You might think that scientists would be objective and self-critical and, to be fair, many of them are, but mostly they are prone to all the weaknesses of humanity at large. Nevertheless, if you were to ask a group of scientists which they thought was the most important, the most fundamental, of all scientific problems, the majority would probably reply that it is to understand the origin of the Universe. But, I hear you say, surely that it is a solved problem and in one sense it is. The big-bang theory starts with an event that occurs at an instant when there is no matter, no space and no time. A colossal release of energy at that 'beginning of everything' spreads out creating matter, space and time. There is some supporting evidence for this model. When we look at distant galaxies we find that they are receding from us at a speed proportional to their distance. If all motions were reversed then in 15 billion (thousand million) years they would all converge at the point where the big bang began — which can be regarded as anywhere since everywhere diverged from the same point at the beginning of time. Before 15 billion years ago there was no time — nothing existed of any kind.

I hope that you understand all that because I certainly do not. It is not that I do not *believe* in the big-bang theory — it is just that I do not really *understand* it. The mathematical model of the big-bang is plain enough and many physicists and astronomers, including myself, can deal with that but I doubt that there are many people on this Earth that really *understand* it. My own test of whether or not I understand something is whether or not I can explain it to others. Sometimes in my teaching career, when I have been preparing a new course, I have suddenly realised that I could not provide a clear explanation for something — the reason being that the topic I *assumed* I understood I did not really understand at all. Fortunately, in the teaching context, by reading and a bit of thought I have been able to deal with my own shortcomings. Nevertheless, I can promise you that I shall not be writing a book on the origin of the Universe.

So, if we now exclude the origin of the Universe and take it as an observational fact that we have matter, space and time, what would most scientists think of as the next most important problem? That surely must be to explain the origin of life. By what process can inanimate matter be transformed even into the most primitive life forms? Is it actually a spontaneous process? We are straying close to religious issues here so I shall go no further in that direction. But, once life exists, even in a primitive form, we do have a theory that *can* readily be understood, Darwin's theory of evolution, which leads us to the higher forms of life, including ourselves. Random genetic mutations occasionally create an individual that has advantages over others in its environment. The principle of 'survival of the fittest' ensures that its newly modified genes flourish and eventually become dominant and by small changes over long periods of time a new species can evolve. There is also some evidence for the occasional rather more rapid creation of new species. Something that Darwin's ideas do not deal with directly is the existence of consciousness, the knowledge of one's own identity and relationship to the outside world. That too is one of the difficult problems of science, although there are those who attempt to explain it in terms of computer technology by asserting that human beings resemble rather complex computers.

The topic of this book is not as important as those mentioned above and in many ways it is much more mundane. The Solar System is a collection of objects — the Sun, planets, satellites etc. — made of ordinary matter — iron, silicates, ices and gases, the properties of which we well understand. The Universe is full of such material so all we have to do is to find a way of transforming it from one state to another. Having said that, although it is not an important problem in a fundamental sense, it is nevertheless quite an interesting one because it turns out to be very difficult. For more than two hundred years scientists have been struggling to find ways of just producing the Sun, planets and satellites, let alone all the other bodies of the Solar System. Part of the problem has been that scientists have tended to concentrate on parts of the system rather than looking at it as a whole. It is as though one was trying to understand the structure and workings of a car by studying just the wheels, the transmission system or the seats. It is only by looking at the whole car that one can understand the relationships of one part to another, how it works and how it was made. The same is true for the Solar System. It is not a collection of disconnected objects bearing little relationship to one another. It is a *system* and it will only be understood by examining it as a system.

The story that I tell goes back a long way — perhaps almost to the beginning of sentient mankind. Starting from the observations that a few points of light wandered around against the background of the stars, a picture emerged of a collection of bodies, connected to the Earth, Moon and Sun, that formed a separate family. Gradually the picture improved until, about three hundred years ago, we not only knew how all the bodies moved relative to each other but also understood the nature of the forces that made them move as they do.

Large telescopes, operating over a range of wavelengths from X-rays to radio waves, together with spacecraft, have given us detailed knowledge of virtually all the members of the solar-system family even to the extent of knowing something about the materials of which they consist. In addition, from Earth-bound observations we have even been able to detect planets around other stars — many of them — so we know that the Solar System is not a unique example of its kind. New knowledge has provided guidance for theoreticians attempting to explain the origin of the Solar System — but it also gives new constraints that their theories need to satisfy.

I began my study in this area, generally known as cosmogony, in 1962, intrigued by the fact that, while there had been many theories put forward, not one had survived close scientific scrutiny. While some of them were superficially attractive they all failed because they contravened some important scientific principle; it is a basic requirement of any theory that *every* aspect of it must be consistent with the science that we know. If a theory explains many things in the Solar System but is in conflict with scientific principles then it is wrong. You can no more have a nearly plausible theory than you can have a nearly pregnant woman. Armed with the knowledge that, since the Solar System exists, there must be some viable theory for its origin I started on what turned out to be a long hard road. There were many dead ends and new beginnings. Ideas arose, seemed promising, failed critical tests and then were abandoned. However, one early basic idea that was the core of what came to be called the Capture Theory survived and evolved. What it evolved into is very different from the starting form but the essential idea is still there. Gradually a picture emerged that seemed to make sense — a good sign — and instead of problems piling up as had been the original experience, it was solutions to problems that seemed to proliferate.

I have already confessed to my lack of deep understanding of the big-bang theory but I *do* understand the nature of planets and related bodies and hence I am prepared to write about the formation of the Solar System. However, in writing a book an author has to consider first the readership for whom it is intended and that was a problem with which I wrestled for some time. A complete deep scientific treatment of all aspects of the various theories would be unintelligible to most non-specialist readers and would just be a reproduction of what is already available in the scientific literature. An alternative approach, in which only verbal descriptions were given throughout, would be more readable but would lack credibility — many theories sound plausible enough when described in hand-waving fashion but wilt under close scientific scrutiny. So, to maximize the readership while maintaining scientific integrity, I have decided on a middle course. Fortunately the level of science needed to deal with most aspects of cosmogony can be understood by anyone with a fairly basic scientific background. The approach that has been adopted is to introduce equations here and there to provide scientific substance together with narrative to explain what they mean. In addition, for those who wish to delve more deeply into the subject, reference will be given to a small selection of books and papers in scientific journals — but I stress, these are *not* essential reading! Hopefully, this text alone will provide an account in a form that should both be understood by the non-expert reader and also be of interest to those with wider knowledge of astronomy or general science.

Prologue: The Dreamer

Gng lay on his back with his head cushioned on a bale of ferns. He was well away from the fire, on the windward side, so that the sky he saw was clear of the sparks that flew high into the air. His stomach was distended with aurochs meat, the product of the successful hunt that he and the other men of the tribe had carried out that day. The women too had done well, with a rich harvest of berries and roots that accompanied the meat in their gargantuan meal. These were the good days. He shivered with apprehension as he thought of the bad days that would soon come. Even with his bearskin cloak and leg covering he would feel the bitter cold. There would be many days when cold and hunger filled his mind to the exclusion of all else. These were times when the wild beasts were as desperate as the members of the tribe and last year two of the children had been taken.

He studied the pattern of lights in the sky that he had come to know so well. His imagination created pictures like those seen in a fire, but the pictures in the sky never changed. There is the snake, there the bear, the waterfall seemed particularly brilliant tonight and the hunter's spear is as clear as ever. Actually there were some occasional changes. Many moons ago he had seen a bright light suddenly appear in the sky. The Moon had been eaten by the night so he could see that this light could make shadows. The brightness had lasted for two moons or so, gradually fading until it could no more be seen. Perhaps there was a great forest in the sky and a fire had raged in it and gradually died away. Such things were part of his experience. There were other changes that were less exciting but that were always happening. As each night passed so the sky lights twisted round, like a leaf on the end of a filament of a spider's web but always in one direction. They all twisted together so the patterns did not change. But, over the course of time, he had seen three lights that slowly moved amongst the others. What were the lights that kept their rigid patterns and what were those that travelled amongst them? Were those travellers like the old rogue males expelled from the herd by a new young dominant bull? No, that did not seem to fit.

He never spoke to the others about what he saw and what he thought. They knew that he was a little different — the name they had given him meant 'dream' or 'dreamer'. But he was a brave hunter and a respected member of the tribe so the difference was tolerated. Once, in a moment of rare tenderness, he had tried to explain to Nid, his woman, about what he saw in the sky. He could not find the words to express his thoughts and she comprehended nothing. She roughly pulled away from him, looked at him with a puzzled and troubled expression and then returned to suckling their latest infant. That was instinct — that she understood.

Gng did not know that he was a very important man. He had observed the night sky and tried to make sense of what he saw in terms of what he knew. He was the first astronomer.

GENERAL BACKGROUND

Chapter 1

Theories Come and Theories Go

It's all kinds of old defunct theories, all sorts of old defunct beliefs, and things like that. It's not that they actually live on in us; they are simply lodged there and we cannot get rid of them.

Hendrik Ibsen (1828–1906), *Ghosts*

1.1. What is Science?

Science is a quest for knowledge and an understanding of the Universe and all that is within it. Individual scientists learn from those that have preceded them and their work guides those that follow. Arguably the greatest scientist who has ever lived, Isaac Newton recognized this debt to his predecessors by saying "If I have seen further it is by standing on the shoulders of giants."

All that Newton discovered is so much the accepted background of scientific endeavour today, at least in astronomy and physics, that what he did may now seem to be obvious and humdrum. Yet, in its day, it was spectacular. It was as though humankind, or at least those who could understand what Newton had done, had a veil moved from before their eyes so that all that was previously obscure was seen with a crystal-like clarity. The forces of nature that caused the Moon to go around the Earth and the Earth to go around the Sun were quantified. Forces that operated in the same way, but with different causes, could explain the way that electric charges attracted or repelled each other and also the behaviour of magnets. While all agree that Newton was a great man and his discovery of the law of gravity was a great discovery, can it be said that it was truth in some absolute sense? Apparently not, because three hundred years later another famous scientist, Albert Einstein, showed that Newton's law of gravitational attraction was just an approximation and, to be *very* precise, one should use the Theory of General Relativity instead. It turns out that Newton's way of describing gravity is good enough for most purposes and the calculations that send spacecraft to distant solar-system bodies with hairline precision use Newton's equations rather than those of Einstein.

The example of gravitation is a good one for portraying one aspect of scientists' attitude to their work. Some of them are purely interested in theoretical matters, in that they just try to understand the way that nature works without necessarily having some practical motive to do so. One of the early deductions from the Theory of General Relativity is that light from a star, passing the edge of the Sun, will be deflected twice as much as would be suggested by Newton's gravitational theory. It was realised that if this could be shown to be true, then General Relativity would get a tremendous boost in credibility. This prediction about the deflection of light was made by Einstein in 1915 while he was working in Berlin during the First World War. The observational confirmation that Einstein was right was made in 1919 by British teams of scientists, led by Sir Arthur Eddington and the Astronomer Royal, Sir Frank Dyson. They travelled to South America and West Africa to make observations during a solar eclipse, when starlight deflected by passage close to the Sun could be seen. The expedition was planned in 1918 while Britain and Germany were on opposite sides of a vicious and destructive war but their scientists could come together in their search for knowledge.

The demonstration that Einstein's prediction was right excited the scientific community, and even members of the general public who realised that something important had happened even if they did not quite know what it was. Although scientifically important, this demonstration was *not* important in making a great impact on everyday life. However, the life that we live today *is* very much shaped by the science that has been done in the last 200 years. Experiments with 'Hertzian waves' in the latter half of the 19th century eventually led to radio, television and the mobile telephone. Einstein's Special

Relativity Theory suggested the idea that matter can be turned into energy (and vice-versa) through the famous equation $E = mc^2$. The world has not been the same since the first large-scale demonstrations of the validity of that equation when atomic bombs were dropped on Hiroshima and Nagasaki in 1945. Curiosity about the way that electrons behave in semiconductor materials led to the electronics revolution that so dominates world economics. In fact, rather oddly, it sometimes seems that curiosity-driven research seems to outperform utility-driven research in terms of the usefulness of the outcome. When, in 1830, Michael Faraday waggled a magnet near a coil of wire and produced an electric current he was not conscious of the fact that he was pioneering a vast worldwide industry for generating electricity.

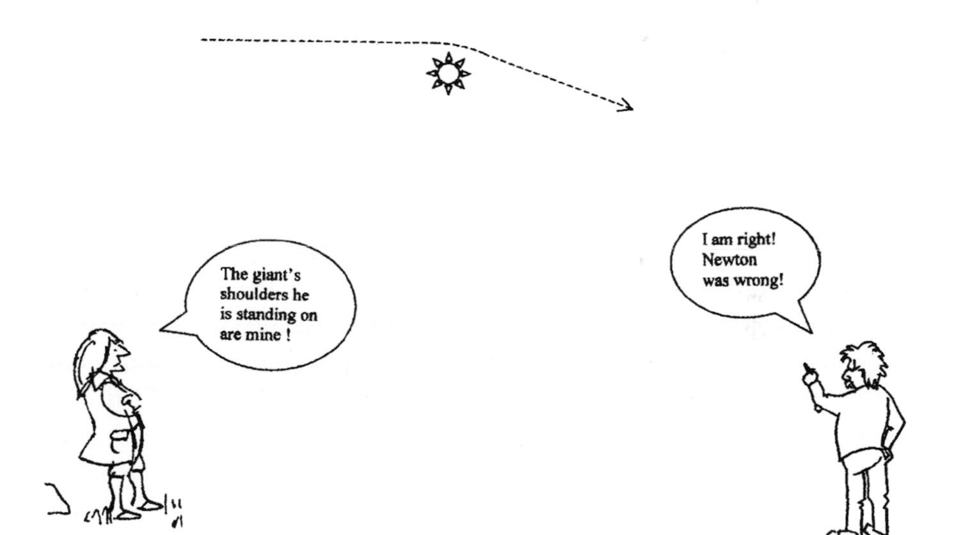

1.2. The Problem of Cosmogony

Although by no stretch of the imagination could one envisage any practical outcome from the deflection of a beam of light passing the Sun, at least it was possible to do the experiment to show that the prediction was true. There are other areas of science where there is limited opportunity for experimentation and one of those is the subject we deal with here — technically referred to as *cosmogony*. The question we address is 'How was the Solar System formed and how has it evolved since it formed?' While there are various ideas in this field, everyone agreed that the Solar System formed some 4,500 million years ago and that, since that time, it has undergone many changes of an irreversible kind. The concept of reversible and irreversible changes is quite fundamental in science. For a reversible change the system could, at least in principle, run backwards so that the past states of the system can be deduced from its present state. For example, we know where the Earth is in its orbit around the Sun and the laws of mechanics that govern its motion. We now imagine that the direction of motion of the Earth reversed so that it retraced its path or, in practice, we do the calculation corresponding to that reversal of motion. This enables us to find exactly where it was at times in the past — but not too far in the past as even the Earth in its motion has undergone some irreversible processes. As an example of an irreversible process, we imagine a large cubical evacuated chamber with a tap at each of its eight corners leading to a cylinder of gas. The tap is turned on at one of the corners and the chamber fills with gas. Once the gas has occupied the whole chamber there is no way of telling from which of the corners the gas came in. The event was irreversible and one cannot make the molecules of the gas reverse their motions and re-enter the cylinder from which they came. For an irreversible change, the past state cannot be deduced from the present state.

If we cannot work backwards to find out how the Solar System began then what *can* we do? The answer is to try various models that are scientifically plausible to see whether or not they can give rise to a system like the Solar System today, or even one that might have evolved to give it. Taking this approach runs the risk that there would be a huge number of models that lead to the Solar System as we know it — but this turns out not to be the case. As we shall see, finding a model that gives anything like the Solar System has proved to be a very difficult exercise.

1.3. New Theories for Old

The history of science is peppered with ideas that have held sway, that were eventually found to be flawed and were then replaced by some new ideas. The lesson to be learnt from this is that no theory can ever be regarded as 'true'. There are two categories of theory — those that are plausible and those that are implausible and therefore probably wrong. Any theory in the first category is a candidate for the second whenever new observations or theoretical analysis throw doubt upon its conclusions. There is no shame in developing a theory that is eventually refuted. Rather, the generation and testing of new ideas must be regarded as an essential part of the process through which scientists gain the knowledge and understanding they seek. The Earth-centred theory of the structure of the Solar System due to Ptolemy, a 2nd Century Alexandrian Greek astronomer (Section 3.5), was a useful model for the 1,400 years of its dominance and it agreed with what was known at the time. People's everyday experience suggested that the Earth was not moving because there was no sensation of movement such as one would have when walking or riding a horse. If the assumption that the Solar System was Earth-centred gave complicated motions of the planets with respect to the Earth then so be it — after all there were no laws of motion known at that time that forbade such complication. When Copernicus introduced his heliocentric (Sun-centred) theory in the 16th century there were still no known laws of motion but the attraction of his idea was that, in terms of the planetary motions, it gave simplicity where the previous theory had given complication. It complied with a philosophical principle, known as *Occam's razor*, enunciated by the 14th century English Franciscan monk, William of Occam. The Latin phrase loosely translates as "If there are several theories that explain the facts then the simplest is to be preferred".

A seeker after knowledge and understanding must be cautious about accepting ideas because they seem 'obvious' and fit in with everyday experience. That, after all, was the basis of Ptolemy's model. Other scientists of great stature have made similar errors.

For example, Newton wrote in his great scientific treatise *Principia* as one of his 'Rules of Science':

"*To the same natural effects, the same causes must be assigned.*"

As an example he gave the light of the Sun and the light of a fire but we now know that these lights have very different causes. The heat of the Sun is produced by nuclear reactions while that of a fire comes from chemical changes produced by ignition. Again, Einstein never accepted quantum mechanics, especially the uncertainty principle, a recipe for defining the fundamental limits of our possible knowledge. An implication of the uncertainty principle is that we can never precisely define the state of the universe at any time and therefore we cannot predict what its future state will be. As Einstein wrote in a letter to Max Born "I, at any rate, am convinced that *He* is not playing with dice." By *He*, Einstein meant God.

The watchword in science is "caution". All claims must be examined critically in the light of current knowledge. Any acceptance must be that of the *plausibility* of an idea since the possibility of new knowledge and understanding to refute it must be kept in mind. We must beware of bandwagons and be prepared to use our own judgements; history tells us that bandwagons do not necessarily travel in the right direction!

Chapter 2

Measuring Atoms and the Universe

. . . si parva licet comonere magnis. . . . "*if one can compare small things with great*". . .

Virgil (70–19 BC)

2.1. Measuring Things in Everyday Life

In life in general, and in science in particular, it is necessary to measure things. If we buy some cheese we expect to know how much it weighs and pay accordingly. If we set out on a journey we want to know how long it is so that we can plan the trip and arrive at the required time — time being another quantity that we measure. In the everyday world the units used for measurement are ones that we can relate to. Different societies have devised different measuring systems — for example the Imperial System, with pounds and feet, devised in Britain and used in a slightly modified form in the USA. However, most societies are now converting to the metric system, one of the great legacies of Napoleonic France. The kilogram (kg) is a very convenient unit of mass. One hundredth of a kilogram is the mass of a one-page letter in its envelope and one hundred kilograms is the mass of a very amply built man. A metre (m) is a length we can readily envisage; one hundredth of a metre, i.e. one centimetre, is roughly the thickness of a finger and one hundred metres is the length of the shortest sprint race in the Olympics. For longer distances the kilometre (km), for example, one thousand metres is a better unit, i.e. the distance from London to Edinburgh is 658 km. Time has a variety of units, although for the scientist the basic unit is the second (s), approximately the time of a heartbeat to put it in familiar terms. The unit of time, at least from when time was first defined, was the day, the time interval between successive noon times, when the Sun is at its zenith. This was divided into hours, minutes and seconds to give a way of defining the time of day with sufficient precision for most human activities and to express periods of time with convenient magnitudes for particular purposes. We would find it difficult to comprehend the time of a 100 m athletics race as 0.0001157 days rather than 10 s and the time taken to cross the Atlantic in a ship as 510,400 seconds rather than 6 days!

Another quantity that needs to be measured in everyday life is temperature. The prevailing temperature indicates what kind of clothing needs to be worn. The baby's bath should be at a comfortable temperature and body temperature can be an important diagnostic indicator of health. In the UK and USA the Fahrenheit scale is still frequently used and understood by the population at large. On this scale the freezing point of water is 32 °F and its boiling point 212 °F. The rather more logical Celsius (centigrade) scale, with the freezing and boiling points of water as 0 °C and 100 °C respectively, is in general use in most of the world and is supplanting the Fahrenheit scale. However, even the Celsius scale is not logical enough for a scientist. Temperature is a measure of the energy of motion of the atoms in a substance. As the temperature of a gas is increased so the gas atoms move around faster. The *Absolute or Kelvin scale* of temperature measurement (the unit, the kelvin, indicated by K) is defined such that the temperature is proportional to the mean energy of their motion. Atoms in a solid are fixed in a rigid arrangement but they do oscillate around some average position and, once again, the temperature is proportional to the mean energy of this vibrational motion. If there is no energy of motion then the temperature is zero on the Kelvin scale[1]. The increments corresponding to one degree are made the same as those of the Celsius scale. This gives 0 K $= -273.2$ °C,

[1] To be precise, even at 0 K there is some residual energy, which physicists call *zero-point energy*. However, we can disregard this in our discussion.

the freezing point of water, $0\,^{\circ}\mathrm{C} = 273.2$ K and the boiling point of water, $100\,^{\circ}\mathrm{C} = 373.2$ K.

2.2. Science and Everyday Life

Simple science that was developed until about 150 years ago was mostly about phenomena that played a part in everyday life. The laws of mechanics are ones that can be understood intuitively because they form part of our experience. Children playing 'catch' with a ball know nothing about the mathematics of parabolic motion but they understand instinctively how it works in practice. A footballer does not have to be a scientist to curl a ball into the corner of the net and the magical performances of great snooker players are based on experience, not scientific qualifications. Few people know about the intrinsic nature of light but everyone knows that you cannot see around corners. Understanding the way that the material world behaves has 'survival value' and so such understanding governs our instinctive behaviour.

2.3. Small Things Beyond Our Ken

One problem in understanding modern science is that it encompasses phenomena that are well outside the experiences of everyday life. We are not conscious of the presence of individual atoms, the size of the Universe does not impinge on our daily lives and the mechanical objects that form our environment do not move at a large fraction of the speed of light. The idea of an atom originated from the 5th Century B.C. Greek, Leucippus. He imagined what would happen if one cut up a piece of matter over and over again making it smaller and smaller. He concluded that eventually an ultimate indivisible piece of matter would remain — and this was the atom. We now know that the atom itself consists of even smaller components. An atom contains *protons* with a positive charge, *neutrons* with no charge but virtually the same mass as a proton and *electrons* with a negative charge, equal in magnitude to that of the proton, but with a tiny fraction of the proton's mass. In some situations, when an atom breaks up, mysterious particles called *neutrinos* are produced which have no charge, possess energy and momentum and have an extremely tiny mass — or, perhaps, no mass at all! We cannot cope with the concept of a neutrino in a framework of everyday experience! There are even smaller particles, *quarks*, from which the atomic sub-particles are made but we shall not go further in that direction. What we have established is that there is the world of very tiny objects, a world that does not directly relate to everyday life.

If we wish to write down the mass of a proton then, in decimal form, it is

$$0.00000000000000000000000000167 \text{ kg},$$

a representation that is almost unintelligible. A slightly better representation is

$$\frac{1.67}{1,000,000,000,000,000,000,000,000,000} \text{ kg}$$

but even this looks very clumsy and is very difficult to decipher. The divisor contains 27 zeros and it represents a product of twenty-seven 10s or, in other words, 10 to the power 27. We have a way of expressing this so that now we can write the mass of a proton as $\frac{1.67}{10^{27}}$ kg or, by a final transformation, 1.67×10^{-27} kg. This is now a far more succinct expression and with experience one can even begin to get a feel of what such a quantity means. In the same notation the mass of an electron is 9.1×10^{-31} kg so that about 1,835 electrons have the same mass as a proton.

Just as for mass, so atomic particles have extremely small sizes. A small atom has a diameter of about 10^{-10} m, which means that a million of them side by side will have the width of a dot over the letter 'i'. A proton is even smaller, 10^{-15} m in diameter so that 100,000 of them will fit across an atom. Because the masses and linear dimensions of atoms and elementary particles are so small compared with the basic units, the kilogram and the metre, atomic and nuclear scientists have devised new units. Thus the *atomic mass unit* (amu) is roughly the mass of a proton but is defined as one-twelfth of the mass of a carbon atom. For length a convenient unit is the *fermi*, which is 10^{-15} m, approximately the size of a proton.

2.4. Measuring Things in the Solar System

Tiny atomic particles are important to scientists and, indeed, to technologists as well since much of the modern communications industry depends on the way that electrons behave. However, they have no obvious part to play in explaining how the Solar System and other planetary systems arose nor are the small dimensions required to describe them relevant in this respect. So now we will move on to the world of the very large — at least by human standards.

The orbits of planets around the Sun are not circles but ellipses. Such an orbit is an oval shape, as shown in Figure 2.1, with the Sun displaced from the centre.

Most planetary orbits are very close to circles, so close that if the orbit of the Earth had been shown in the figure then the departure from a circle could barely have been detected visually. It is clear that as the planet goes round the Sun so its distance from the Sun is constantly changing. The point marked *perihelion* is when it is closest to the Sun and the point marked *aphelion* (pronounced afelion) is when it is furthest away. The terms perihelion and aphelion can also be used for the actual distances from the Sun.

The extent to which an ellipse departs from being a circle is measured by a quantity called its *eccentricity* denoted by e. One way of describing eccentricity is

$$e = \frac{\text{aphelion} - \text{perihelion}}{\text{aphelion} + \text{perihelion}} \tag{2.1}$$

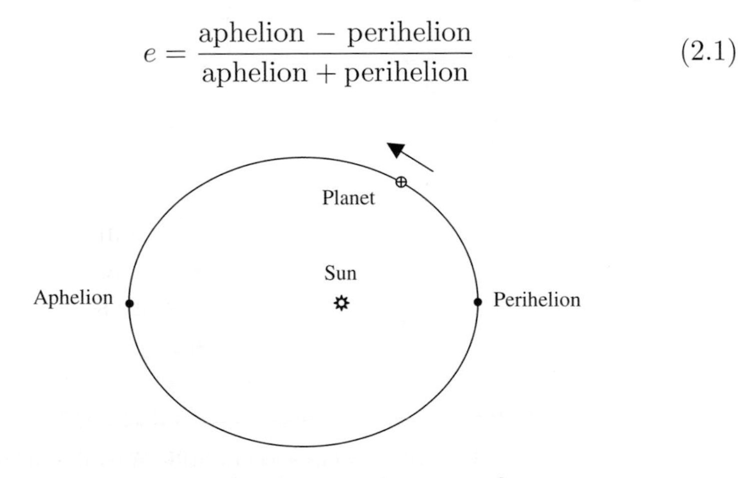

Figure 2.1. An elliptical planetary orbit.

For a circular orbit e will be zero since the planet is always at the same distance from the Sun and the aphelion and perihelion will be equal. For an ellipse e can be zero up to anything less than 1. For the Earth the aphelion is 1.521×10^8 km and the perihelion is 1.471×10^8 km which, put into (2.1), gives $e = 0.017$. Shown in Figure 2.2 is a selection of ellipses with a range of eccentricities together with the position of the Sun if the ellipses represented planetary orbits.

The eccentricity of an orbit describes its shape but not its size. The length of the line joining the aphelion to perihelion is called the *major axis* of the ellipse and half that distance, the *semi-major axis*, gives the size of the ellipse. It is approximately the average distance between the planet and the Sun, with the average taken over a complete orbit.

An important distance in the Solar System is the semi-major axis of the Earth's orbit, which is 1.496×10^8 km. This distance is called the *Astronomical Unit* (au) and is a very useful unit for measurements in the Solar System so that, for example, the semi-major axis of Jupiter's orbit is 5.2 au.

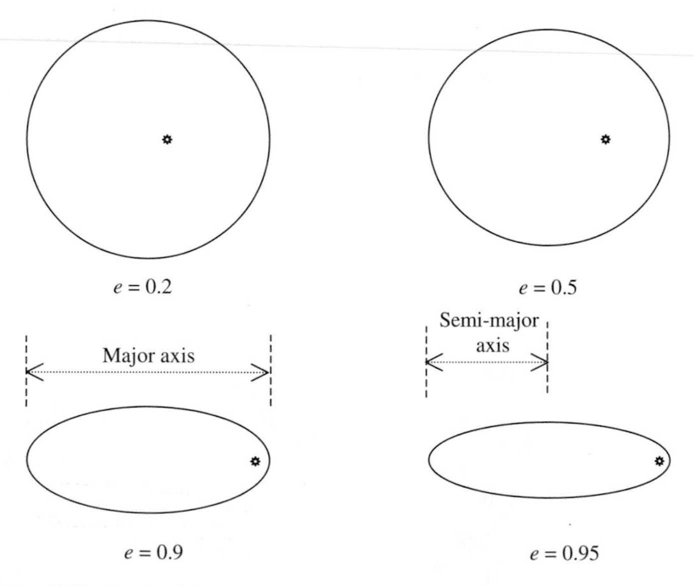

Figure 2.2. Elliptical orbits with various eccentricities. Also shown are the major axis and the semi-major axis.

The characteristics of the Earth and its orbit are also used to define mass and time in many solar-system contexts. The Earth mass is 5.975×10^{24} kg; the most massive planet, Jupiter, has mass 317.8 Earth units while Mercury, the innermost planet has mass 0.0553 Earth units. Similarly the time taken for the Earth to go round the Sun, the year, is a convenient unit for measuring the orbital periods of other planets — 11.86 years for Jupiter and 0.241 years for Mercury.

2.5. Large Things Beyond Our Ken

We have dealt with the kinds of measurements we need for discussing aspects of the Solar System. But, in terms of the Universe at large the Solar System is a tiny and insignificant entity. To measure distances in the Universe, the metre, or even the Astronomical Unit, is far too small to be convenient. Instead we use the Light Year (Ly), the distance travelled by light in a year, as a convenient unit of measurement. Light travels at 300,000 km s^{-1} (kilometres per second) and there are 3.16×10^7 seconds in a year so

$$1\,\mathrm{Ly} = 300{,}000 \times 3.16 \times 10^7\,\mathrm{km} = 9.5 \times 10^{12}\,\mathrm{km}.$$

The Sun is a member of the Milky Way galaxy, an island universe containing 10^{11} stars. Figure 2.3, a picture of a galaxy with the unromantic name NGC 6744, which is very similar to the Milky Way, shows what it is like. The distance across it is about 100,000 Ly. In the Universe that we can detect with the most powerful telescopes,

Figure 2.3. The galaxy NGC 6744 — very much like our own Milky Way galaxy.

there are about 10^{11} galaxies. The nearest large one, the Andromeda galaxy is close at hand, just 3 million Ly away, while the furthest objects we can see are some 10^{10} Ly away. The Universe is expanding and these very distant objects are rushing away from us at a large fraction of the speed of light — which is fascinating — but another story.

ENLIGHTENMENT

Chapter 3

Greek Offerings

Timeo Danaos — et dona ferentis. "*I fear the Greeks, even when they bring gifts*".

Virgil (70–19 BC)

3.1. Even Before the Ancient Greeks

Three to four thousand years ago, while in most parts of Europe people lived in tribal communities or minor fiefdoms, great civilizations had grown up in Babylonia, Egypt, India and China. The very essence of a civilized community is that it has strong central control and the individuals within it have varied roles that contribute to the whole structure of society. Astronomers, who studied the heavens for very practical reasons, played an important role in these early civilizations. The times for the sowing and harvesting of crops needed to be known and religious ceremonies had to take place at particular times of the year. Such needs dictated the establishment of a calendar, which, in its turn, required the study of the motions of the Moon and the stars. Another strong motivation for studying the heavens was to assist in astrological predictions. It was strongly believed in many early societies (and by some individuals even today!) that the fate of humankind is bound up in the motions of heavenly bodies. Eclipses and comets were the harbingers of great events and so it was important to know when they were coming and what they signified.

What the astronomers saw was great regularity and predictability in the heavens. They did not really comprehend the true nature of what they observed but they built up empirical rules that gave them powers of prediction — for example, of solar eclipses. There was just one phenomenon that departed from the smooth regular behaviour of all the other bodies they saw. These were the wandering stars, five of them, of which three made occasional looping motions in the sky while two others could only be seen close to the Sun when it was below the horizon, just after sunset or before dawn.

These early astronomers were humans and therefore they were curious — as are all humans. But we know little of any speculations they may have made about the nature of the objects they were observing. For that, we must move forward to ancient Greeks, whose offerings we need not fear — just admire.

3.2. Plato and Aristotle

The Greeks were the first philosophers, in the archaic sense of that word which is synonymous with scientists today. Even in the twenty-first century the Physics department of the University of Edinburgh is named the *Department of Natural Philosophy*. However, the Greeks indulged in a curious mixture of philosophical concepts in the modern sense and of science as we would understand it. Thus they had a concept of beauty and perfection and they felt that natural objects should conform to these concepts. Plato (c. 428–348 BC), who founded a great philosophical school, the Academy, thought that heavenly bodies had to be spherical because spheres were the *perfect* shape. For the same reason of perfection he thought that all the motions of heavenly bodies had to be circles. Well, whatever the reasoning he got the right answer for shapes, at least for all but very small astronomical bodies of which he would have been unaware in any case.

Greek philosophy did not confine itself to astronomical matters but here we shall just highlight aspects of their thinking that advanced understanding about astronomy, in particular relating to the Solar System. A pupil of Plato, Aristotle (384–322 BC), had ideas ranging over the complete ambit of human thought. He conceived of a universe structured in three parts. The outer bound of the universe was a huge spherical shell in which the stars resided. Closer in was a region occupied by the Sun and the planets and finally, right at the centre, was the Earth and the Moon. The placement of the Earth at the centre of the universe, with everything else moving around it, is the most intuitive of beliefs since we have only to study the heavens to see that everything moves round the Earth — or so it seems.

3.3. Aristarchus — A Man Ahead of his Time

Aristarchus of Samos (310–230 BC) was an Alexandrian Greek with a rather untypical attitude to science in that he was prepared to do experiments. The Greeks had the practice of believing that if they came to a conclusion by the power of thought then that conclusion had to be true. For example, they had reasoned that heavy objects fall faster than light objects — which is certainly not true but they did not think it necessary to test the hypothesis.

Aristarchus was interested in knowing how far the Moon and Sun were from the Earth and he designed some experiments to find out. He understood that the Moon was illuminated by the Sun and from geometrical considerations this meant that when half of the Moon was seen illuminated, then the angle Earth-Moon-Sun had to be a right angle (Figure 3.1).

Aristarchus tried to measure the angle θ shown in the figure. This angle would not give the actual distances of the Sun and the Moon from Earth but it would give the ratio of the distances of the two

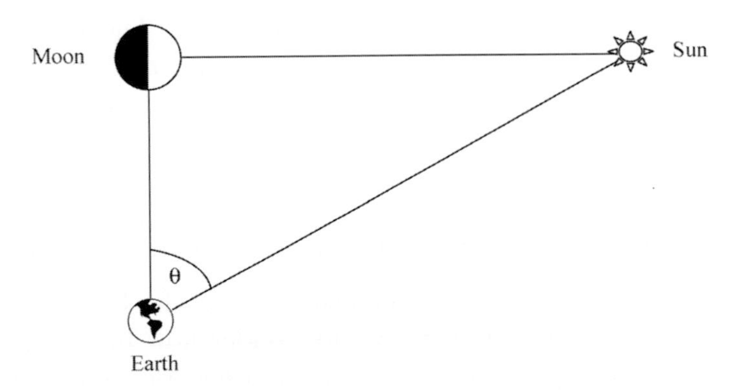

Figure 3.1. The Sun-Earth-Moon arrangement at half-Moon.

bodies. What Aristarchus did not know was that the distance of the Sun from Earth is much greater than the distance of the Moon so that the angle θ is only a little less than 90°. This meant that, since small differences from 90° made a large difference to the estimated ratio, θ had to be measured very accurately. Aristarchus estimated the ratio of the distances to be 19 ± 1, much less than the true value of 390. The idea for the measurement was sound but the technology of the time was inadequate.

In addition to finding the relative distances of the Moon and the Sun, Aristarchus also made estimates of the relative sizes of the Earth, Moon and Sun. Since at the time of a total eclipse of the Sun the Moon just covers the Sun's disk it followed that the sizes of the two bodies were just proportional to their distances. Hence, according to Aristarchus, the Sun's radius was just 19 ± 1 times that of the Moon. To find the size of the Earth relative to the Moon, Aristarchus used the phenomenon of a lunar eclipse when the Earth's shadow passes over the Moon's surface. He noted that the arc of the Earth's shadow was of considerably larger radius than that of the Moon itself and he estimated the ratio of the radii as about three. Actually the radius of the Earth is nearly four times that of the Moon but the idea was established that the Earth was considerably bigger than the Moon and the Sun much bigger than both the other two bodies.

Aristarchus had another idea that was well ahead of its time. He proposed that the Sun and the stars were fixed in position and that the Earth moved in a circular path around the Sun. The idea that the Earth moved round the Sun would not be raised again for another 1,800 years and the general acceptance of that proposition would take even longer.

3.4. Eratosthenes — The Man who Measured the Earth

We have seen that the Greeks understood that the Earth is spherical in shape, so naturally they would wonder how big it was. The answer to that question was provided by another Alexandrian Greek, Eratosthenes (276–195 BC). It was well known by local people that at mid-day on the day of the summer solstice, when the Sun was at its furthest northern position, it shone straight down a well at Syene (near modern Aswan). This meant that the Sun was at the *zenith*, i.e. directly overhead. Alexandria is due north of Syene so the Sun was not directly overhead but shone down at an angle to the vertical.

In Figure 3.2 the Sun's rays point straight down the well at S, the location of Syene. In Alexandria, located at A, a tall tower throws a shadow the length of which, relative to the height of the tower, gives the angle θ. This angle is the difference in latitude between Alexandria and Syene. Eratosthenes estimated this angle to be 1/50th of a complete rotation so that the circumference of the Earth was just 50 times the distance between Syene and Alexandria. Distances in those days were measured by paces; there were people trained to make a standard pace just as soldiers have always done. A Roman soldier made 1,000 paces to a Roman mile (a word based on the Latin *millia* meaning 'thousands'), where a *pace* corresponded to two steps in the Roman interpretation. In British infantry regiments a standard pace (single step) is 30 inches, except for light infantry regiments where it is 27 inches; if Roman soldiers had used a 30 inch step then

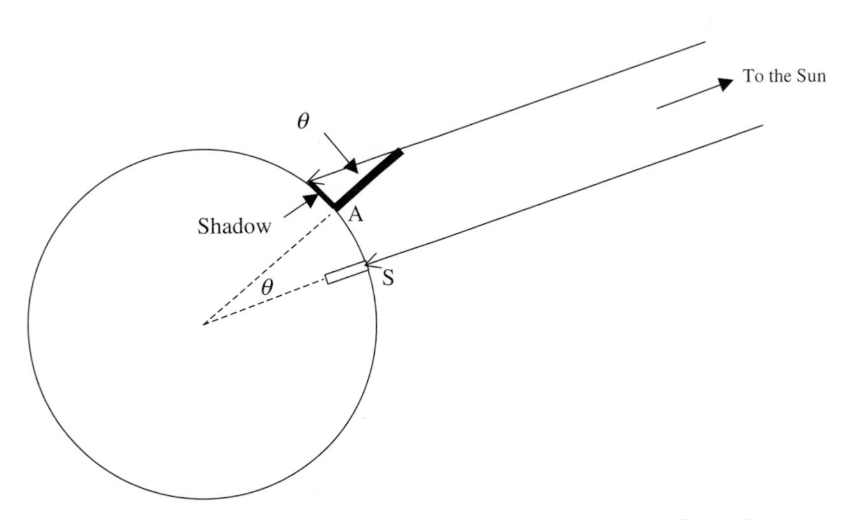

Figure 3.2. Light from the Sun shining straight down a well at Syene but casting a shadow from a tower in Alexandria.

the Roman mile would have been some 5% less than a British mile. The professional pacer found the distance from Syene to Alexandria to be some 5,000 stadia so Eratosthenes estimated the circumference of the Earth to be 250,000 stadia. The stadium is believed to have been about 180 m, giving the radius of the Earth as 7,200 km according to Eratosthenes — about 13% larger than the actual value of just under 6,400 km.

3.5. Ptolemy and the Geocentric Solar System

Ptolemy

We all sense the world from the viewpoint of our own position in it, which is a statement of the obvious. Here you see a tree, there a house and above both of them a bird is circling in the sky. Since you have the imagination that all humans possess, you can deduce what the scene would appear to be from the viewpoint of a distant hill or perhaps of the bird — but it is an effort. What is true for one person is true for humankind as a whole. We see the universe from a stable unmoving platform, the Earth, and all other bodies, the Sun, the Moon, the planets and the stars move around it. For the ancient Greeks to have thought that the Earth was moving would have been almost impossible. Motion was something that was felt by the mover, walking or being transported by horse, wagon or ship. Clearly the Earth did not move.

This geocentric, i.e. Earth-centred, view of the Universe was supported by the motions of the Moon, Sun and stars that all moved in smooth circular motions in the sky. These motions agreed with a proposition put forward by Aristotle that the motions of all heavenly bodies were in circles at a constant speed. The exceptions to this harmonious picture were the planets that generally moved in an easterly direction but also, periodically, made great looping motions in the sky (Figure 3.3). Indeed this is why they had been given the name planets, derived from the Greek word meaning 'wanderers'.

It was yet another Alexandrian Greek, the astronomer Ptolemy (100–170 AD) who brought order to the motions of the planets. Since simple circular motion at a constant speed could not provide an explanation for their motion, he devised a scheme involving two circular motions at a constant speed — illustrated in Figure 3.4. The first motion in a circle at constant speed was of a point called the *deferent*. The planet then moved round the deferent in a circle at constant speed, this path being called an *epicycle*. This model

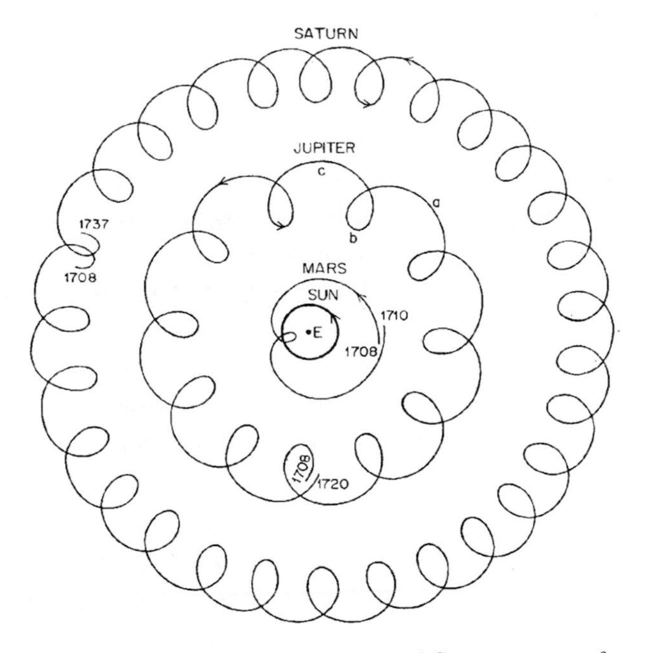

Figure 3.3. Motions of Mars, Jupiter and Saturn as seen from Earth.

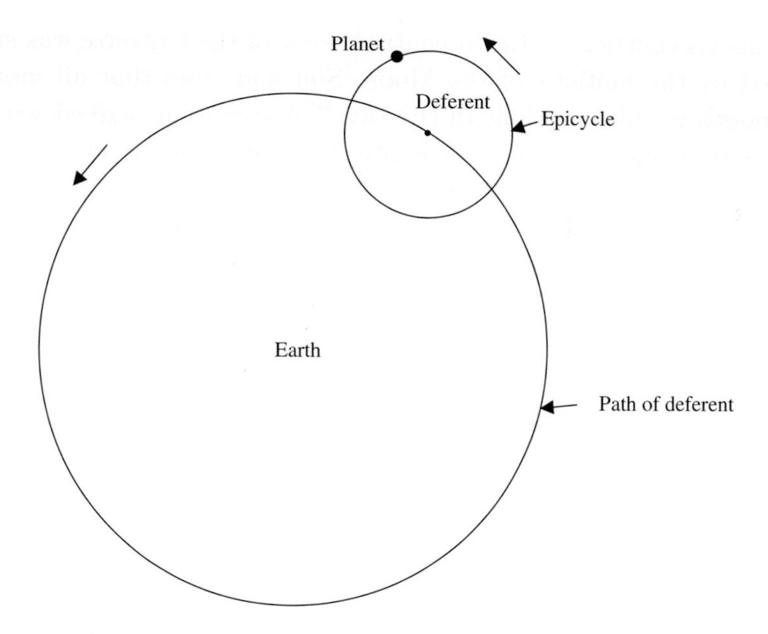

Figure 3.4. Ptolemy's description of planetary motion.

reproduced the motions of the planets quite well, certainly to within the accuracy that the planetary motions were known at that time. There were also some rules that the motions had to obey. Since Venus and Mercury are always seen close to the Sun, the deferents of these planets always had to be on the Earth-Sun line. Again, to explain the motions of the planets outside the Earth's orbit, the *superior planets*, the line joining the deferent to the planet always had to be parallel to the Earth-Sun line. These were complicated rules but it has to be remembered that there were no laws of mechanics known at that time and Ptolemy's model at least had the virtue that it was a systematic description of planetary motions. Indeed this was the model that was to be generally accepted for the next 1,400 years.

Chapter 4

The Shoulders of Giants

We are like dwarfs on the shoulders of giants, so that we can see more than they, and things at a greater distance, not by virtue of any sharpness of sight on our part, or any physical distinction, but because we are carried up and raised high by their giant size.

Bernard of Chartres (c. 1130)

4.1. The Refugees

The beginning of Roman Empire can be traced back to 510 BC. Over several hundred years the Romans occupied large parts of Europe, North Africa and the Middle-East. In general they were not noted for their scientific or philosophical contributions but they were skilful administrators and they established a legal code — Roman Law — a novel institution in many of their conquered territories. They were extremely competent civil engineers, especially skilled in the building of aqueducts, bridges and roads, and the routes of the roads they built are still followed by many modern highways. They were also adept in military technology and they fully exploited the resources, both in material and manpower, of the regions they conquered. Thus legions recruited in Syria manned Hadrian's Wall, along the northern border of their British territories, and legions recruited in Britain served in other parts of the Empire.

The Empire was often beset with turmoil and intrigue as various factions and individuals sought to become dominant. Eventually, in 395 AD the Empire split into two parts, the Western Roman Empire centred on Ravenna (since Rome came under frequent barbarian attack) and the Eastern Roman Empire with capital Constantinople. Both Empires had Christianity as the state religion; Constantine, who died in York in 337 AD, was the first Christian emperor. The last emperor of the Western Roman Empire was overthrown by a German chieftain in 476 AD and thereafter the heritage of Rome was confined to the Eastern Empire.

After the birth of Islam in the seventh century AD, the Eastern Empire was greatly influenced by the poetry, science and philosophy of the surrounding Arab nations. However, the Turks occupied Constantinople in 1453 and many Christian scholars fled eastwards to an Italy that, by this time, existed as collection of individual states within all of which there was a strong influence of the Roman Catholic Church. The renaissance, a scientific, artistic and literary revival, had begun in Italy in the 14th century but this received a great fillip from the arrival of cultured and educated refugees from the east. This is the background for a new surge in astronomical understanding, especially about the Solar System.

4.2. Nicolaus Copernicus and a Heliocentric Solar System

Nicolaus Copernicus (1473–1543) was a Polish cleric who spent some time as a professor of mathematics and astronomy in Rome and also practiced medicine. In the 15th century, and indeed for some time thereafter, it was not uncommon for learned men to extend their studies over many apparently disparate fields, very unlike the situation today. The vast explosion of knowledge in the last two centuries has, perforce, led to increasing specialization so that, for example, the average physicist would not be completely familiar with all topics in physics outside his or her immediate field of interest.

Copernicus became interested in the motions of the planets and he had available translations of Ptolemy's *Almagest*, the 13 books containing the accumulated astronomical knowledge of the 2nd century AD. Copernicus constructed tables of planetary motions and, from the currently available better observational data, he concluded that Ptolemy's description of geocentric planetary motion was unsatisfactory apart from being rather complicated, which is a negative feature of any theory, then as now.

Copernicus

Using his improved observations he developed a heliocentric model in which the planets all circled the Sun. However, like Ptolemy before him, he was reluctant to abandon the concept that planetary motions had to be based on circular motion at a constant speed. The observations required that the angular speed of planets in their motions around the Sun had to vary slightly and he satisfied this requirement by having the centre of the circular motion displaced from the Sun. When the planet was closer to the Sun then it would have a greater *angular* speed, which is just what was needed. Even with this modification there were still discrepancies between the motions of his model planets and the observations, so he introduced small epicycles to improve the agreement although the required epicycles were tiny compared with those postulated by Ptolemy.

The church was aware of his astronomical work and was either supportive or, more likely, was perhaps just disinterested in what he was doing. Copernicus delayed the publication of his work for about 20 years and it eventually appeared in the treatise *De Revolutionibus Orbium Coelestium* which he dedicated to Pope Paul III. Just after the book was published Copernicus died — it was said that he received the printed copy on his deathbed. Anyway, he was not to see the storm that his work was destined to stir up by the end of the century.

4.3. Tycho Brahe — the Man with a Golden Nose

Tycho Brahe (1546–1601) came from a noble family with close connections to the Danish royal family. He quarrelled with a fellow student at his university and in the ensuing duel his nose was sliced off — after which he wore a false nose made of gold and silver. He was a quick-tempered and arrogant man and these attributes would greatly influence his subsequent career.

Tycho was interested in astronomy from an early age. In 1572, a very bright supernova appeared and he observed it over an eighteen-month period. The king, Frederick II, was very impressed by the young Tycho and offered to support his work on a grand scale. He made available to him the island of Hven, between Denmark and Sweden, and the resources to construct and run a large observatory. This observatory, called Uraniborg, was equipped with line-of-sight instruments based on very large quadrants for measuring angles, designed by Tycho himself. The engraving in Figure 4.1 shows one of these quadrants mounted against a wall of the observatory. The picture painted on the wall depicts Tycho, but the man himself can be seen, towards the centre of the right-hand margin of the engraving, taking an observation. One assistant is recording the time from a clock while another assistant, seated at the table, is writing down the observations.

At the time that Tycho was making his observations the heliocentric model of the Solar System, as proposed by Copernicus, was well known and it might be thought that its clear superiority over the Ptolemaic system would have recommended it to all practicing astronomers. This was not so. The general acceptance of the geocentric model was so well established that there was a considerable resistance to change. Lest we be critical of this conservatism of a bygone age we should note that, even today, there is reluctance to abandon old ideas in favour of new ones — no matter how strong the evidence. Tycho himself adopted a hybrid model in which all the

Figure 4.1. Tycho Brahe's quadrant.

planets other than the Earth orbited the Sun but with the Sun and the Moon orbiting the Earth. Actually, this correctly gave all the relative motions of the bodies, which was all one needed in the absence of any scientific theory to explain how they all moved. For that, one had to wait more than one hundred years.

Tycho ran into trouble when his patron, King Frederick II, died in 1588. The residents of the island of Hven were Tycho's tenants and he treated them very badly, a consequence of his arrogance and disdain for those he considered his inferiors. Frederick's successor, his son Christian, was less tolerant of Tycho's bad behaviour than his father had been and, in addition, he felt that Tycho took for granted the generous support he had been given and for which he showed very little appreciation. Eventually Tycho had to leave Hven and in 1599 he joined the court of Rudolph II in Prague as the Imperial Mathematician. This was a fortunate move. While in Prague he compiled tables of planetary motions, based on his years of accurate observations, assisted by a very able young man, Johannes Kepler. It was this young man who was to take the next significant step in increasing understanding of the nature of the Solar System.

4.4. Johannes Kepler — A Mathematical Genius

Johannes Kepler (1571–1630) was born into a poor family but, due to a rather enlightened government in Württemberg where he was born, he was able to receive an education. He first became interested in the Copernican model when he was a student at Tübingen and some time later, when he taught mathematics at Graz, he wrote a book, *Mysterium Cosmographicum*, in support of the heliocentric theory. While he supported the general idea that all planets, including the Earth, orbited the Sun, he was not happy with the description given by Copernicus that required small epicycles. He felt instinctively that the orbits should be a simple curve of some kind; unlike Copernicus he was not obsessed by the idea, originating with Plato, that all orbits should be based on circular motions.

Kepler

Kepler sent a copy of his book to Tycho and this began a correspondence that eventually, in 1599, led to Tycho inviting Kepler to be his assistant at a new observatory that was being constructed at Benatek. Kepler readily accepted the offer — the thought of being able to analyse the superb data collected by Tycho was more than he could resist. In the event, he was disappointed as Tycho gave him only limited access to his data but when Tycho died the data became his and he was able to begin his great project of analysing exactly how planets moved. The key to this work was the motion of Mars, the well-observed planet that most departed from circular motion.

After eight years of effort, Kepler discovered the first two of his three laws of planetary motion; the third law took another nine years to be formulated. These laws are:

(1) Planets move in elliptical orbits with the Sun at one focus.
(2) The radius vector sweeps out equal areas in equal time.
(3) The square of the period is proportional to the cube of the mean distance.

The first two laws, together with the determination of the shape of the orbit of Mars, were published by Kepler in *Astronomica Nova* in 1609 and the third law was given in his book *Harmonica Mundi* in 1619.

In retrospect, it seems rather surprising that Kepler took so long in finding that the orbits were ellipses. He was a skilled geometrician and the ellipse was a well-known geometrical shape.

Although Kepler is best known for his laws of planetary motion he also made contributions in many other areas of science. He was the first to propose that tides were due to the action of the Moon — although his great contemporary, Galileo, disagreed. He also knew that the Sun spun about its axis, a motion that he described in *Astronomica Nova*. He was interested in understanding how a telescope worked so he investigated the way that images were formed, including image formation by the eye. He also gave practical rules for designing eyeglasses for long and short sightedness. His optical work was published in two books, *Dioptrice* (a term still used today) and *Astronomia Pars Optica*. He also understood how two eyes, giving slightly different views, could give depth to perception and suggested how the distances of stars could be found by observing them from points separated by a diameter in the Earth's orbit. The attempts he made to measure the distances of stars were unsuccessful because the instruments he used were not good enough but the principle is sound and has been used in modern times with great success.

4.5. Galileo Galilei — Observation versus Faith

Galileo Galilei (1562–1642), a contemporary of Kepler, was appointed as professor of mathematics in Pisa at the age of 25 and also taught at Padua University for 18 years. He was interested in mechanics and also in planetary motion. Kepler sent him a copy of his *Epitome of the Copernican Astronomy* and in response Galileo indicated that he too favoured the heliocentric model.

Galileo

In 1600, an event occurred that was to have important repercussions for Galileo. An Italian Renaissance philosopher and Dominican monk, Giordano Bruno, put forward the proposition that other stars were like the Sun and hence would have accompanying planets occupied by other races of men. This threw down a distinct challenge to the Church and its doctrine of the central role of mankind, moulded in God's image and with a unique relationship to the deity. Bruno was brought before the Inquisition but refused to retract his views and paid for his steadfastness, or obstinacy, with his life. The source of the trouble, as far as the church was concerned, was the Copernican model and its view of *De Revolutionibus* underwent a complete reversal. In place of support, or tolerance, the work was added to the *Index Librorum Prohibitorum*, the list of forbidden books and it was joined on the list by Kepler's writings. In case it be thought that the Catholic Church was alone in being against the heliocentric theory it should be stressed that it was also anathema to the Lutheran Church that was dominant in the north of Europe. However, the Lutheran Church did not have such power and influence in northern states as the Catholic Church wielded in the south, which meant that Kepler could pursue his ideas unmolested. This was not so for Galileo.

In 1608, following an invention of some spectacle makers in Flanders, Galileo made a telescope. He was encouraged to do this by the Venetian Senate. With even a low-power telescope, ships could be seen approaching the harbour while still two hours' sailing away, which had both commercial and military value. However, Galileo saw the telescope as a powerful tool for astronomy and he used it to great effect. He saw lunar mountains and made estimates of their heights by the length of the shadows they cast. He also discovered four large satellites orbiting the planet Jupiter — Io, Europa, Ganymede and Callisto — now known collectively in his honour as the Galilean satellites. Seeing these satellites around Jupiter, like a miniature version of the Solar System, reinforced Galileo's belief in the essential correctness of the Copernican model. Another important observation in 1610 was of the planet Saturn that he saw with a pair of appendages looking like ears. We now know that what he was looking at were Saturn's rings but the quality of his telescope was too poor to show a clear image. Galileo was astonished when in 1612 the appendages disappeared; the rings were then edge-on to his line of sight but he did not know that.

None of these observations had relevance to the controversy concerning the geocentric and heliocentric alternative models of the Solar System. But there was one observation made by Galileo that had direct relevance and that was of the changing phases of Venus.

According to Ptolemy's model the deferent of Venus always had to be on the Earth-Sun line because only then would Venus always be seen at a small angle from the Sun.

According to this picture, illustrated in Figure 4.2, as seen from Earth it would always be the rear of Venus that was illuminated by the Sun so that only a part of the edge of Venus would appear lit up and thus it could only be seen as a crescent.

The Copernican model gives a different outcome — shown in Figure 4.3. Both Venus and the Earth are now orbiting the Sun, with Venus in the inner orbit. Sometimes Venus is between the Earth and the Sun and Venus is seen in a crescent phase. When this happens Venus is close to the Earth so the crescent is large. At other times Venus and the Earth are on the opposite sides of the Sun. In this case a whole circular face of Venus is illuminated and a 'full' Venus is seen. When that occurs Venus is at its greatest distance from the Earth and Venus appears small.

Galileo's observations showed clearly that the Copernican theory was the correct one but he was a devout man and the Church had set its face against the heliocentric model — so what could he do?

Galileo decided to try to convince the world at large of the correctness of the Copernican theory without unduly offending the Church.

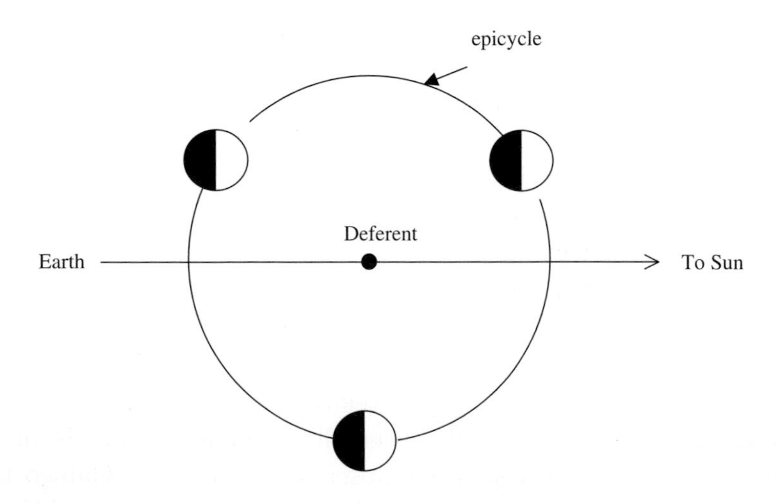

Figure 4.2. Venus as seen from Earth according to Ptolemy's model.

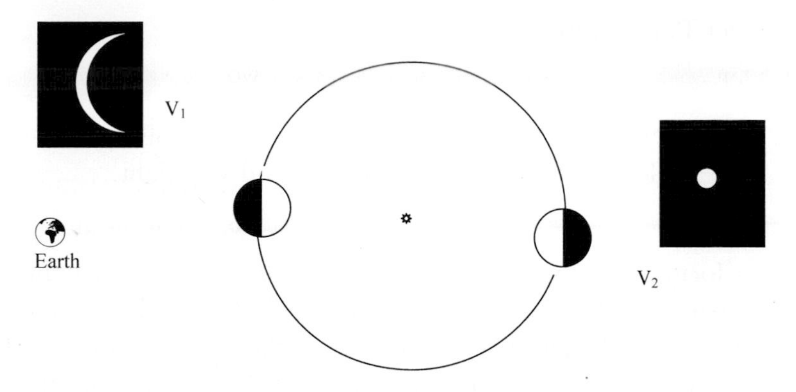

Figure 4.3. Venus as seen from Earth according to the Copernicus model.

In 1632 he wrote a book, *Dialogue on Two World Systems*, in which two individuals, Simplicius and Salvatio, argued the respective merits of the geocentric and heliocentric models. The advocate of the geocentric model, Simplicius, as his name almost suggests, seemed simple-minded compared with the lucid and compelling Salvatio. The device of trying to protect himself by appearing to present an impartial argument did not work. The Pope, Urban III, was furious and Galileo was ordered to appear before the Inquisition. He humbly recanted and proclaimed that he sincerely believed that the Earth was the centre around which all other bodies moved. There is a story, probably apocryphal, that after his recantation he muttered *Eppur si muove* — but still it moves.

Because of his eminence Galileo was not subjected to ill-treatment nor was he imprisoned but he was under virtual house arrest for the rest of his life. His public humiliation was a sad affair but it was probably the last time that the new scientific method of deduction based on observation was challenged on an irrational basis.

4.6. Isaac Newton — and All was Light

On Christmas day in the year that Galileo died, the greatest scientific genius of all time, Isaac Newton (1642–1727) was born. The poet

Alexander Pope wrote

> Nature, and Nature's laws lay hid in night:
> God said, *Let Newton be!* And all was light.

Newton formulated 'Newton's laws of Motion', he studied light and various aspects of hydrostatics and hydrodynamics and he developed the branch of mathematics which he called *fluxions* but is now called *calculus*. His greatest contribution to astronomy was the formulation of the inverse-square law of gravitation that states that "the gravitational force between two point masses is proportional to the product of the masses divided by the square of the distance between them". His laws of mechanics specified that a body could only change its speed or direction if had a force acting on it. Since the Moon in its orbit constantly changes direction it must have a constant force acting on it and the Earth was the obvious source of that force. By this kind of argument, and perhaps watching apples fall off trees, he was led to his law of gravitational force. He was able to demonstrate that Kepler's three laws of planetary motion followed from the inverse-square law of gravitational attraction, in particular that the orbits of planetary bodies round the Sun or satellites around planets were ellipses.

Newton

Newton described his theory of gravitation, and much else he had worked on, in a major publication, the *Principia*, which was published in 1687. It took 15 years to write and indeed he had to be persuaded to publish it by his good friend Edmund Halley, whose name is associated with a well-known comet. Halley even paid for the cost of the publication. It can truly be said that Newton's work represented the greatest watershed in science, in particular in understanding planetary dynamics. From Newton onwards the essential structure of the Solar System was clearly established and clearly understood.

Newton was a somewhat irascible and quarrelsome individual who made many scientific enemies. Robert Hooke, an eminent contemporary scientist, claimed to have considered an inverse-square law for gravitational attraction before Newton had and to this claim Newton reacted with characteristic vehemence. More serious was his quarrel with the German mathematician Leibnitz over who had first invented calculus. The truth of the matter is certainly that they had done so independently and, in fact, it is Leibnitz's mathematical notion that we commonly use today.

Not all of Newton's scientific activities were fruitful. He spent a great deal of time on alchemy, an attempt to change base metals into gold. But, for all his genius, he was a child of his time and alchemy was then regarded as a sensible branch of science — although looked down on by many of Newton's contemporaries. His last years were spent as Master of the Royal Mint. It was characteristic of the man that he pursued counterfeiters relentlessly, many of whom suffered imprisonment and death thereby. The greatest of all scientists, maybe, but not the greatest of all men!

THE SOLAR SYSTEM: FEATURES AND PROBLEMS

Chapter 5

A Voyage of Discovery to the Solar System

I travelled among unknown men.

William Wordworth (1770–1850)

5.1. Travelling Towards the Solar System

There is a well-known saying — "you can't see the wood for the trees" which implies that if you are in the middle of a wood, your view is so obstructed by the surrounding trees that you have little chance of appreciating the composition of the wood as a whole. The situation is not quite as bad in viewing the Solar System from the vantage point of the Earth but, even so, the best way to understand the structure of the Solar System would be to approach it from a distance and gradually see features in greater and greater detail. One would first see the gross structure of the system and then, with closer approach, the fine structure as smaller objects could be discerned. To simulate this way of looking at the Solar System we shall describe what would be reported by a hypothetical space traveller coming from some planet orbiting a distant star. However, to be able to understand his report we shall assume that our space traveller has monitored Earth communications for a long period and knows the names of our planets, our units of measurement and our terminology for Solar System objects.

5.2. Approaching the Solar System

My name is Regayov and I come from the planet Arret that orbits the star Los. I am approaching the star, known to earthlings as the

Sun. It is an average star of temperature of about 6,000 K about half way through its main-sequence lifetime so that in 5,000 million years time it will become a red giant.

Regayov

The first planet I detect is Jupiter, a fairly large one with a diameter of about one tenth of that of the Sun (Figure 5.1). It is a swirling ball, mainly of hydrogen and helium gas, but within it there is almost certainly a rock and metal core with a few times the mass of Earth.

Figure 5.1. The largest planet, Jupiter. The swirls on its surface are storms. The large one, at the bottom of the picture, the Red Spot, has persisted for hundreds of years.

Then, as I get closer, other smaller planets came into view. These are Saturn, Uranus and Neptune and from this distance it seems that the Sun had a family of just four planets. Saturn seems like a smaller version of Jupiter with about one-third of its mass. Uranus and Neptune are quite similar to each other, with masses of about one-twentieth of that of Jupiter, more or less, and they are quite a bit smaller because they contain considerable amounts of denser materials, in particular water, ammonia and methane.

As I approach more closely, I am able to detect another set of orbiting bodies much closer to the Sun. Eventually I ascertain that there are four of them, small rocky bodies, and it is from the third one out, the Earth, that life signals emanate. The innermost planet, Mercury, is small, about one twentieth of the mass of the Earth and completely devoid of an atmosphere. The side that happens to face the Sun at any time is so hot that some metals can melt while the other side is extremely cold. The next planet out is Venus, somewhat smaller and less massive than the Earth but not greatly so. It is covered by a very thick atmosphere one hundred times as thick as that on Earth, consisting mainly of carbon dioxide. The surface of Venus is incredibly hot, some 730 K, much higher than would be suggested just by its distance from the Sun. This is due to a *greenhouse effect* where short wavelength radiation coming from the hot Sun is able to penetrate the atmosphere to the surface and heat it but where longer wavelength radiation from the cooler surface has much more difficulty in escaping. No life form can possibly exist on Venus.

Earth, shown in Figure 5.2, has white clouds made of water vapour and its fairly thin atmosphere mostly consists of nitrogen but with about 19% oxygen that is used for life combustion. Its ambient temperature is mostly at the lower end of the range between which water freezes and boils, although large incursions into the freezing region can occur, especially near the poles. Earth's orbit around the Sun is nearly circular and the relationship of its orbital and spin periods gives a year of just over 365 days. Earth's spin axis is tilted which gives strong seasonal effects on temperatures throughout the year. Earth is home to a vast variety of life forms. The one we call

Figure 5.2. Earth, the largest of the inner rocky planets and the one with life. About 70% of the surface is covered with water, seen as blue in the picture.

earthlings dominate the planet on account of its much greater intelligence — poor though by our standards.

The outermost of the inner planets is called Mars. It is quite small and has about one-ninth the mass of Earth. It is red in colour due to the presence of iron oxides on its surface. It has a very thin atmosphere, consisting mostly of carbon dioxide. The tilt of its spin axis is similar to that of the Earth so this gives seasonal variation of temperatures in the northern and southern hemispheres. This is manifested by frozen ice-caps that advance and retreat with the different seasons. These have a permanent component of water ice, the variable component being frozen carbon dioxide that evaporates and condenses as the seasons change.

There are clear signs, e.g. old dried-up river beds, that Mars once had water on its surface. There are also many extinct volcanoes — one of them, called Olympus Mons, being the largest in the Solar System.

There are many other smaller bodies orbiting the Sun which I shall mention later. One of these small bodies, called Pluto, is regarded by the earthlings as a planet — but then they seem to be very confused about what constitutes a planet. It has a very elliptical orbit so that it can approach closer to the Sun than Neptune does — although the main part of its orbit is well outside that of Neptune. It is tiny, with

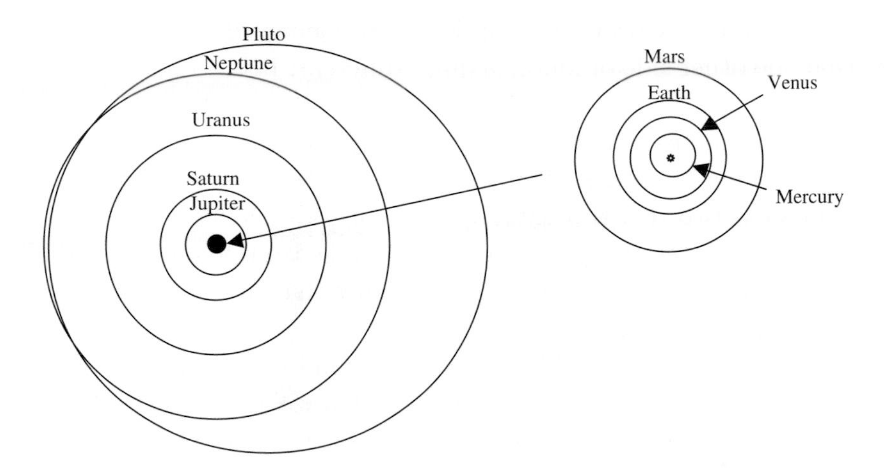

Figure 5.3. The planetary orbits. The orbits of the terrestrial planets are contained in the solid black circle.

about one-twentieth of the mass of Mercury. Although it is not really a planet it does have three satellites, one quite large relative to the planet and the other two very tiny. There are many other satellites in the Solar System but before I describe them, here in Figure 5.3 is a picture giving an impression of the orbits of the planets — as one of which I include Pluto in deference to the earthlings's categorization.

I should say here that all the planets orbit around the Sun in the same sense — that is anticlockwise as seen from the north.

5.3. Most of the Planets have Satellites

I have already mentioned a satellite in relation to the pseudo-planet, Pluto. Actually all the other planets except Mercury and Venus have satellites. One of the most interesting of these satellites is the one orbiting Earth, called Moon, which is quite large in relation to its parent body, having more than one-quarter of its radius and about one-eightieth of its mass. In fact Moon has five times the mass of Pluto! It is a rocky body without an atmosphere. Because of tidal forces one face of the Moon is constantly turned towards Earth and that side shows large basins full of basaltic material that must have come from below the surface in a molten state. These basins, called *mare*, cover a large proportion of the Earth-facing side. The odd thing is that the other side of Moon is quite different. It is all mountainous highland regions with only a few tiny mare basins. It is as though the two sides have had a different history and that difference must say something about its origin.

Mars has two very tiny satellites, Phobos and Deimos, which orbit close to the planet. They very much resemble asteroids, which I shall describe later, and they are probably bodies that have been captured by Mars at some time.

Jupiter has many satellites, most of them quite small, but it has four large ones known as Io, Europa, Ganymede and Callisto. Io, the closest of the large satellites to Jupiter, has active volcanoes. Because Europa, the next one out, has exactly twice the orbital period of Io, these two satellites always come close together at the same points of their orbits. For this reason the gravitational effect of Europa repeatedly nudges Io at the same part of its orbit, which makes the orbit of Io slightly non-circular. Thus the distance of Io from Jupiter alternately becomes greater and smaller as does Jupiter's tidal stretching force on Io. This variable stretching of Io pumps heat into it, as happens with repeated bending and unbending of a piece of metal, and it is this heat that drives the volcanoes. Europa, which is covered in ice, may also have heat generated in it that has melted some of the ice below the surface so that the surface ice may be floating on an ocean. Earthlings speculate that there may be life forms in this ocean.

The outermost large satellites, Ganymede and Callisto, have such low densities that they are not just covered with ice but must have a large fraction of ice in their composition. Ganymede is actually bigger than the planet Mercury and Callisto (Figure 5.4) is only a little smaller.

The remaining satellites are small, most of them being very far from the planet. An outermost group of satellites orbit in a retrograde sense, i.e. clockwise as seen from the north. It seems probable that these small satellites, and another group of small satellites further in, are captured bodies.

Saturn has many satellites, but only one that rivals in size and mass the large satellites of Jupiter. This is Titan that, like Ganymede,

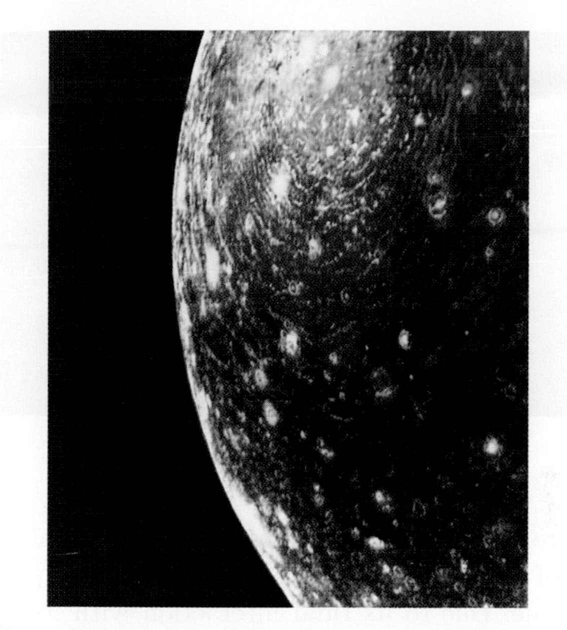

Figure 5.4. Part of the surface of Callisto. It shows a circular impact feature called Valhalla. A large body struck the surface, melted the ice and waves travelled outwards like ripples in a pond. The ripples froze to give the appearance shown.

is larger than the planet Mercury. Its main claim to fame is that it has a considerable nitrogen atmosphere that is actually thicker than that of Earth, although it is a much smaller body. The remaining satellites vary in size from moderate to small. They all orbit Saturn in the usual way except the outermost one, Phoebe, which is retrograde.

Saturn has another type of companion that is most impressive, a ring system consisting of a vast number of solid bodies forming a planar disk. This feature, shown in Figure 5.5, can be seen from far away — indeed, I noticed it as soon as I reached the outer bounds of the Solar System. On closer inspection I found that all the other large outer planets have ring systems but they are flimsy and difficult to see compared with that of Saturn.

The satellites of Uranus are all much smaller than the larger ones of Jupiter and Saturn and have no particular features of note, although some of them have rather interesting broken-up surfaces. On the other hand two of the many satellites of Neptune are full of

Figure 5.5. Saturn and its ring system.

interest. There is a fairly large satellite, Triton, with about one-third of the mass of the Moon, which is in a fairly close *retrograde* orbit around Neptune. Due to its tidal interaction with Neptune, Triton is slowly spiralling inwards and will eventually be broken up by the tidal pull of Neptune and bits of it will fall into the planet. There is another fairly small satellite, Nereid, that is on a very elongated orbit that takes it as close as 1,300,000 km and as far as 9,500,000 km from Neptune. There is a temptation to link these peculiar satellite features with Pluto, whose orbit takes it just within the orbit of Neptune. The orbital periods of Pluto and Neptune are in the ratio 3 : 2 and at the present time the orbits of Neptune and Pluto are so synchronized that they can never get close together — but this may not always have been true. Some event in the past could link Neptune, Pluto and Triton — and perhaps even Pluto's satellites.

5.4. Other Small Bodies

We travellers know that the space between stars is occupied by small bodies but so sparsely that they represent no real threat. There have been reports of near misses but never a collision. As I approach the Sun to within about 40,000 au there is an increase in the number density of small bodies — not high enough to be a real hazard, you understand, but very much greater than in the depths of space. These bodies, mostly a few kilometres across, consist of silicate with a high content of volatile material, mainly frozen compounds of hydrogen, oxygen, carbon and nitrogen. The numbers of these bodies, some of which were many tens of kilometres across, steadily increase and then decrease again as the orbit of Neptune is approached. From time-to-time one of these distant bodies is nudged by a passing star, or by some other influence, into a path that takes it close to the Sun. Earthlings call these objects *comets* (Figure 5.6). When they approach the Sun the volatile materials vaporize and produce a tail consisting of gas and dust which is quite visible.

The inner boundary of the icy bodies is in the region just beyond Neptune and some of the larger bodies in this region have been detected from Earth. This region is known as the Kuiper Belt, named after the astronomer who predicted it before it was discovered. One body that might be considered to be part of the Kuiper Belt is Eris, a spherical body which is somewhat larger than Pluto. It moves on a very eccentric orbit, of eccentricity 0.849, with perihelion 76 au. Its description as the 'tenth planet' is a matter of dispute by earthlings who, as I have previously mentioned, seem very confused about what constitutes a planet. This is due to the fact that they do not really understand how planets form in the first place.

As I approach close enough to the Sun to see the inner planets it seems to me from the progression of orbital radii that something is missing between Mars and Jupiter. When I approach much closer

Figure 5.6. A typical comet.

Figure 5.7. A typical asteroid.

I observe that this gap region is heavily populated by small objects called asteroids (Figure 5.7). There are many thousands of them and they sometimes collide with each other or with the planets. The fragments from collisions between asteroids are very numerous and they fall on the planets in large numbers. Earthlings call them meteorites and they are highly treasured as objects of study. It has been found that some are predominantly iron, others silicates of various kinds and a small proportion are mixtures of iron and silicates.

Meteorites are mostly far too small and rare to pose much risk of causing extensive damage on Earth. However, there is evidence that at intervals of tens or hundreds of millions of years asteroids do strike the Earth and other planets. In the case of the Earth this can so change conditions on the planet that a large proportion of the life on it can be destroyed. There is evidence that 65 million years ago such an event eliminated a whole class of non-mammalian creatures called dinosaurs, which had dominated life on Earth for more than one hundred million years. The present dominant species, earthlings, originated less than one million years ago but, on present evidence, will destroy itself long before the next asteroid arrives.

Chapter 6

The Problem to be Solved

Mankind always sets itself only such problems that it can solve...

Karl Marx (1818–1883), *A Critique of Political Economy*

6.1. Knowledge and Time

It is sometimes difficult to appreciate the state of knowledge that existed even in the earlier years of our grandparents, forebears that most young people have actual contact with. I knew my great-grandmother, who was born in 1845 and died when I was fifteen years old. The Crimean War ended when she was eleven years old, a war in which more soldiers died of disease than in battle. These were mostly diseases that can now be cured almost routinely by the application of antibiotics. At the beginning of the 21st century the great scourges of cancer and AIDS killed millions every year. My primary-school grandchildren, who nonchalantly operate mouses (or even mice!) on their computers, will probably look back on my generation as one of pathetic ignorance in which people died of cancer and AIDS because of the lack of some routine treatment. That is the way of the world. Knowledge advances with time and we can only properly judge the achievements of any generation in terms of the knowledge base within which it operates.

What is true for medicine is also true for knowledge about astronomical matters. Uranus was discovered in 1781, the first asteroid was detected in 1801, Neptune discovered in 1846 and Pluto in 1930.

Any theoretician working on the problem of the origin of the Solar System does so within the framework of the knowledge of his time. He is in the position of a detective faced with the task of solving a crime for which there is no witness. His clues are the observations about the present state of the Solar System and he will try to deduce some sequence of events that will explain them. He must use his imagination, but he cannot let his imagination run riot because there are certain scientific laws that must be obeyed. When he comes up with a proposed solution to the mystery, he must present it to a remorseless judge and jury, the scientific community, that will never say that he is right but might, reluctantly, concede that his analysis is plausible — but only that. There is no verdict that stands for all time. If new evidence comes forward that throws doubt on the solution then it will be discarded. One engaged in this area of investigation is in the position that no matter how many facts his or her theory explains, if it runs counter to just *one* important fact or physical principle, then the theory is wrong.

With this picture in mind we look at a range of theories proposed for the origin of the Solar System up to about the year 1960. These theories had very limited objectives — just to explain the existence of the major bodies in the system — and there was no attempt to explain any of the finer details, such as the presence of asteroids and comets.

Before describing these earlier ideas, we briefly describe the gross structure of the Solar System in a little more detail than hitherto and hence compile a list of essential requirements for a basic theory produced before 1960.

6.2. Very Basic Requirements for a Solution

Seen at its very crudest level, the Solar System consists of the Sun and a planetary family. The planets are all in orbit around the Sun in the same *direct* sense (anticlockwise as seen from the north) and their orbits are approximately coplanar — the greatest deviations being Pluto at 17° to the mean plane and Mercury at 7°. The planets are divided into two groups, the terrestrial planets occupying the inner

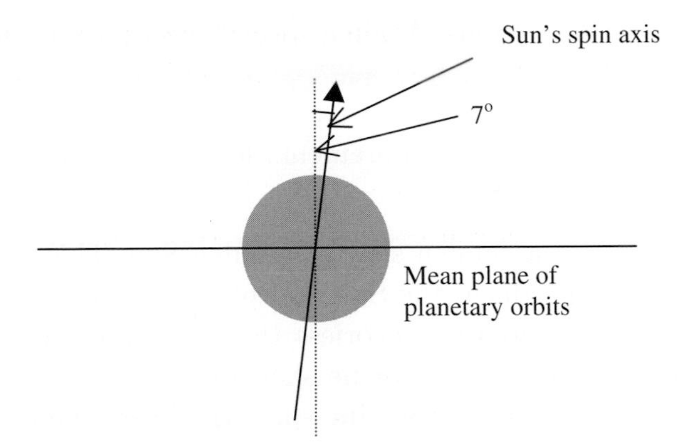

Figure 6.1. The relationship of the Sun's spin axis to the mean plane of planetary orbits.

part of the system and the major planets further out. Pluto seems anomalous in being a very small body in the outermost region.

The Sun spins very slowly, about once in 30 days and, as illustrated in Figure 6.1, the spin axis is inclined at 7° to the perpendicular to the mean plane of the planets. The Sun has about 700 times the combined mass of the planets.

There are satellite families associated with all the major planets. For Jupiter, Saturn and Uranus the innermost satellites are all in direct near-circular orbits that are almost precisely in the equatorial planes of the spinning planets. Neptune has two satellites[1], Triton which is in a close *retrograde* circular orbit and Nereid in a very extended and highly eccentric orbit. The satellites associated with the terrestrial planets are somewhat anomalous. The Moon is a very large satellite in relation to the Earth and its relationship to the Earth might be explained by some special event rather than by the normal processes that formed the Solar System. The two small satellites of Mars are very close to the planet and they have characteristics of size (and appearance — but this was not known in 1960) that are

[1] Actually there are many other satellites discovered by spacecraft but here we reflect the state of knowledge up to about 1960.

similar to those of asteroids. Again a special event for their formation is likely.

Given this gross description of the Solar System, the following are suggested as essential requirements for any plausible theory of the origin of the Solar System:

(1) It must give or assume a slowly spinning Sun. This means that if the theory produces the Sun and planets as part of the same process (called monistic theories) then the Sun must either be produced spinning slowly or be shown to go through some subsequent process that slows its spin. However, there are some theories that assume an already existing slowly spinning Sun (dualistic theories) and just concentrate on forming planets. This just sweeps the problem of the slowly spinning Sun under the carpet — but it is still there!
(2) It must produce planets in more-or-less coplanar direct orbits. If a theory could give the two different types of planets — terrestrial and major — in the correct general locations then that would be a bonus mark.
(3) It should give satellite families, at least for the major planets.
(4) It should explain the 7° tilt of the solar spin axis. This is unlikely to be a coincidence since there is only a 1 in 270 probability of having a tilt of 7° or less just by chance.

Of course, much more was known about the Solar System in 1960 than has been described above but for now we ignore this as it was not relevant to the objectives of the work up to that time.

Chapter 7

The French Connection

The French are wiser than they seem...

Francis Bacon (1561–1626), *Of Seeming Wise*

7.1. Some Early Theoretical and Observational Developments

The first idea with some kind of scientific basis about the origin of the Solar System was due to René Descartes (1596–1650), a French philosopher and mathematician. At that time, before Newton, although it was known how the planets moved there was no real understanding of the mechanics that governed the behaviour of the Solar System. The Descartes model was rather vague and qualitative and based on observations of fluid motion. Descartes postulated that space was filled with a universal fluid, of unspecified nature, that formed vortices around stars. Eddies in the vortices then produced planets and a smaller system of vortices around planets went on to produce satellite systems. Although it had no physical basis the theory did deal with the problem of the planarity of the system and the common direction of orbital motion of the planets and satellites. Another idea, that was later developed, was due to Immanuel Kant (1724–1804) who described a process by which a cloud of dust would take on a disk-like form, although the arguments he used would be unacceptable today.

By the end of the eighteenth century there were advances in telescope technology that enabled better observations to be made. William Herschel (1738–1822), the discoverer of Uranus in 1781, the

Descartes Kant

first new planet to be discovered since antiquity, had constructed a magnificent telescope, the "40-foot reflector" that was installed in his garden in Bath and was the best of its kind for more than half a century. With this instrument he observed fuzzy patches of light that he speculated were "island universes" — the first intimation of the existence of other galaxies. However, in 1791 he also observed a nebulosity that seemed to be centred on a single star and in his various writings there can be found suggestions that this was somehow linked to planet formation.

William Herschel

7.2. Laplace and his Spinning Cloud

The theoretical work of Descartes and Kant and Herschel's observations were probably influential in the formulation of the first substantial scientifically-based theory of the origin of the Solar System proposed by the French scientist, Pierre Simon Laplace (1749–1827).

Laplace

Laplace started his model for the formation of the Solar System with a hot spinning cloud of gas and dust. As it cooled it also collapsed and as it collapsed, it spun faster. This increase in the spin rate is associated with a physical quantity called *angular momentum*, which is so important that it merits a brief description. We consider a body rotating about some axis with a spin rate that, for our purposes, we can think of in terms of 'revolutions per second'. We now think of the body as consisting of a vast number of tiny bodies, that we can call *elements*, each of which has very small linear extent. In Figure 7.1 we schematically show the large body and one of the elements.

The angular momentum of the element, J, is defined as the product of

$$\text{mass } (m) \times \text{ square of the distance } (r) \text{ from the spin axis} \\ \times \text{ spin rate } (\omega)$$

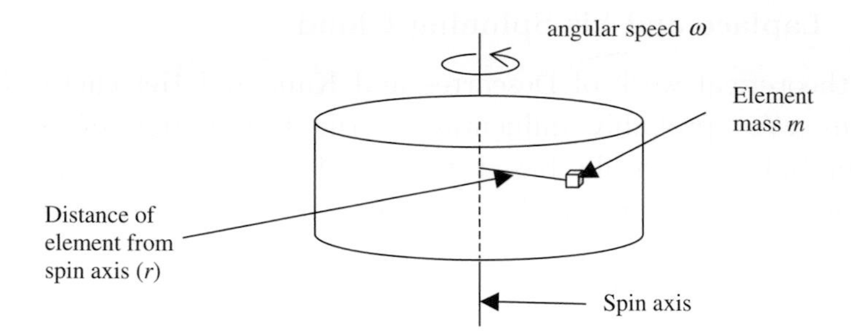

Figure 7.1. An element of a spinning body.

or, in symbols,

$$J = mr^2\omega. \tag{7.1}$$

The quantities, added up for all the elements that constitute the body, give its total angular momentum. Angular momentum has the important property that for any body that is completely isolated, i.e. not acted on by forces from outside, the angular momentum remains constant. Actually, this is a matter of common observation. In exhibition ice skating or dancing, the skater will slowly twirl on her skates with arms extended gracefully outwards. She then brings her arms to her side and spins at a much faster rate. What has happened is that by bringing in her arms she has reduced their distance from the spin axis. To compensate for this and to keep angular momentum constant the spin rate must therefore increase.

Just as for the ice skater, when the spinning cloud shrinks, then, to keep its angular momentum constant, it must spin faster, and the more it shrinks the faster it spins. As it spins so it becomes flattened along the spin axis until eventually it takes on a lenticular form, the shape of a glass lens with sharp edges (Figure 7.2). Eventually it is spinning so fast that material at the edge of the lenticular shape gets left behind in the equatorial plane. Laplace postulated that what was left behind would be in the form of a set of rings and that, eventually, the central core of material collapsed inwards to form the Sun. In each ring, material would gradually accumulate by mutual gravitational attraction until it formed a planet. Before the planet

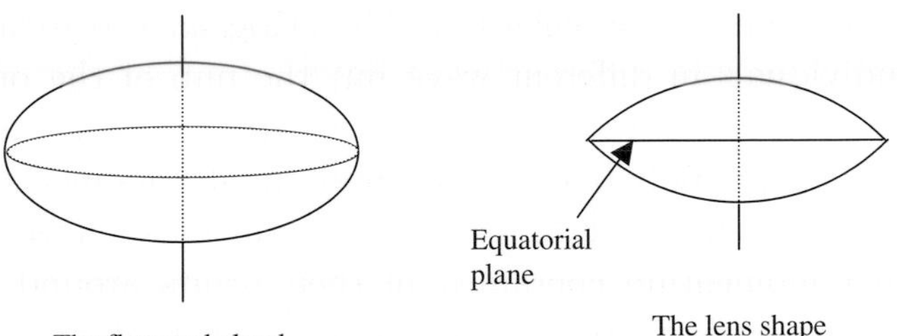

Figure 7.2. Stages in the collapse of a spinning cloud.

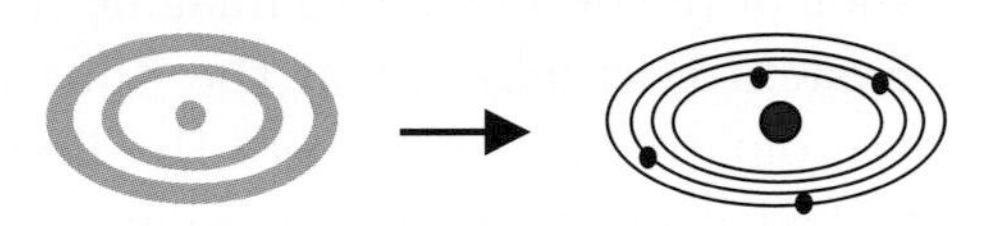

Figure 7.3. The formation of, first, annular gaseous rings and, finally, one planet in each ring.

formed it would be in the form of a gaseous collapsing sphere and a small scale version of the process that formed planets around the Sun would give satellites around each planet. These final stages of planet and satellite formation are shown in Figure 7.3.

The theory had an attractive simplicity. A large gaseous sphere, for which there was some observational evidence, spontaneously gave rise to the Sun, planets and satellites by a fairly straightforward mechanical process. Laplace presented his idea in 1796 in a book *Exposition du système du monde.* Initially the theory received great support that lasted throughout the nineteenth century. However, during that period various objections began to be made and gradually doubts began to creep in about the validity of the theory.

7.3. The Problem with a Spinning Cloud

The Laplace model satisfied three of the four basic requirements for a successful theory given in the last chapter. It certainly attempted to explain the existence of the planets and their satellites and gave a planar system, but it made no reference at all to the spin of the Sun. By the middle of the nineteenth century there were many criticisms of Laplace's model but the one that had the greatest impact involved the concept of angular momentum. The criticism was expressed by several individuals in different ways but the nub of the problem is the fact that the Sun, with 99.86% of the total mass of the system contains in its spin only 0.5% of the total angular momentum of the system. The planets, with 0.14% of the total mass, have 99.5% of the angular momentum contained in their orbits around the Sun. There did not seem to be any reasonable way that the mass and angular momentum of the original nebula could become partitioned in that way.

Another expression of the problem is to imagine that all the material of the Solar System (virtually all in the Sun) is expanded to occupy all the space out to Neptune where the first ring was supposed to have detached itself. Since the angular momentum would have to be the same then as it is now (remember it is constant in an isolated system) then the spin period would be about 3 million years, much too slow for any ring to have detached itself. Alternatively, if when the Mercury ring detached itself the nebula stretched out to Mercury then it would have been spinning with Mercury's orbital period, 88 days. By the time the remaining material had shrunk to the size of the Sun it would have been spinning with a period of about 17 minutes. Actually the Sun could not spin at that rate because it would fly apart.

There were attempts to rescue the Laplace nebula theory by postulating unlikely distributions of material in the original nebula. For example, a French astronomer, Eduard Roche, considered an original nebula in which most of the material was concentrated at the centre with very little on the outside. This certainly improved the situation with respect to the mass-angular momentum distribution (it was still pretty bad, though) but it meant that there was too little material in the outer part of the nebula to produce planets at all! The various arguments against Laplace's nebula model were so strong that by the beginning of the twentieth century support for it had virtually ceased and the scientific community was receptive to some new idea.

Chapter 8

American Catherine-Wheels

You're hidden in a cloud of crimson Catherine-wheels.

Christopher Fry (1907–2005), *The Lady's not for Burning*

8.1. Spirals in the Sky

The American geologist, Thomas Chrowder Chamberlin (1843–1928) examined the Laplace model in some detail, in particular in respect of its implications for the structure and history of the Earth. Eventually he began to lose faith in the Laplace nebula idea although not necessarily in the idea of the formation of planets from diffuse material. As a result of his background in geology he came to the conclusion that the properties of the Earth would be best explained by an accumulation of solid bodies — which could come about by condensation from some kind of nebula. He soon became associated with a much younger man, Forest Ray Moulton (1872–1952), who was an astronomer and mathematician and whose skills and expertise complemented his own.

In 1900, new observations were being made that were to greatly influence the ideas of Chamberlin and Moulton. These were of spiral nebulae (Figure 8.1) that we now know are complete galaxies, very much like the Milky-way galaxy within which our Solar System resides, but their true nature was then unknown. It was assumed that they were part of our own galaxy and the image was interpreted as being a star surrounded by material that could potentially produce planets. Chamberlin and Moulton became convinced that this could

Figure 8.1. A spiral nebula.

provide the scenario for a successful theory of the origin of the Solar System.

8.2. Making a Catherine-Wheel

The first idea considered by Moulton and Chamberlin for producing a spiral nebula was as a result of a collision between two neighbouring nebulae. It was soon realised that this would be a very unlikely event and could not plausibly explain the numbers of observed spiral nebulae. Shortly afterwards they concentrated their interest on solar prominences, large eruptions of matter from the surface of the Sun (Figure 8.2). These eruptions do not lead to a loss of material, as

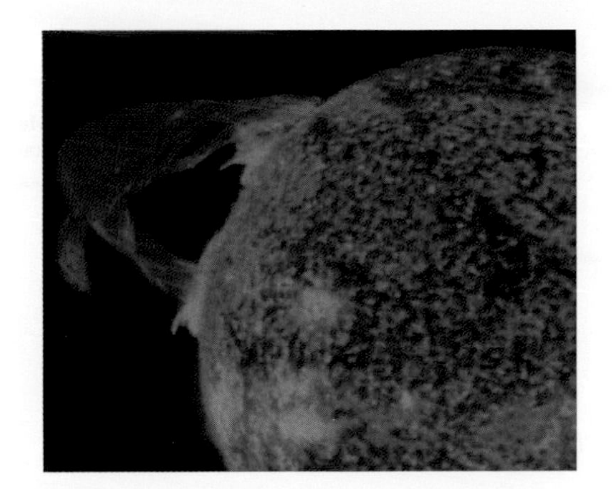

Figure 8.2. A solar prominence.

they loop back to rejoin the Sun, but at the height of a large prominence, material is much more loosely bound to the Sun than when it is part of the surface material.

Chamberlin and Moulton conceived the idea that at a time when the Sun was particularly active, a massive star happened to pass close enough to the Sun to pull the solar prominences outwards and leave them in orbit. The tidal effects that caused this to happen would operate most strongly in the direction towards the star and also outwards on the opposite side of the Sun. This is the way tides work. The high tide in the sea caused by the Moon occurs not only in that part of the Earth closest to the Moon but also in the opposite side of the Earth furthest from the Moon.

In Figure 8.3 we see a schematic representation of two streams of matter that have been pulled out of the Sun. The bunching of density in the streams is due to the assumption that the Sun loses material in a spasmodic fashion. The dense regions cool rapidly and then liquid and solid objects that Chamberlin and Moulton called *planetesimals* would form. Planetesimals would then accumulate to form the planets and smaller collections of planetesimals would go into orbit around the planets to form satellite systems.

The theory is dualistic in the sense that it assumes a pre-existing slowly rotating Sun before the process of forming the planets begins. However, Chamberlin and Moulton did propose an explanation of the

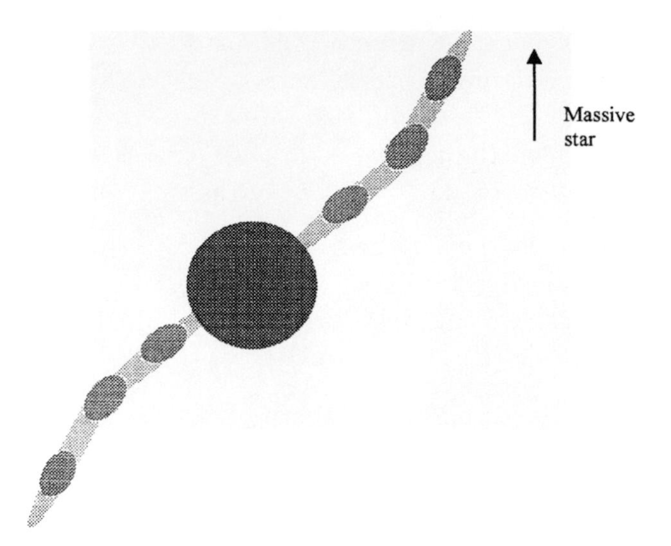

Figure 8.3. Condensation in a solar prominence drawn out by a passing star.

7° tilt of the solar spin axis as being due to the passage of the massive star that pulled the prominences out of the Sun's equatorial plane. Although the theory has many *ad-hoc* features, it does seemingly satisfy the basic conditions for a plausible theory.

8.3. Objections to the Chamberlin–Moulton Theory

The Chamberlin-Moulton theory did not receive much support outside the United States, mainly because it was very qualitative with hardly any mathematical analysis to support it. A German astonomer, Friedrich Nölke, put forward some particularly telling arguments against it in 1908. Some of these were:

(1) Stellar interactions would be too rare to explain the observed numbers of spiral nebulae.

(2) To pull out the prominences, the star would have to be at a distance such that the tidal effects on the two sides of the Sun would be very asymmetrical. By contrast, spiral nebulae are quite symmetrical.

(3) If the inner part of the spiral corresponded to Mercury and the outer part to Neptune, then because of the very different orbital

periods of the planets, the spirals would distort very quickly and the probability of observing them as they are actually seen would be extremely small.

(4) Even though the passing star would be fairly close it would not be close enough to give the 7° tilt of the solar spin axis.

All debate about the validity of the Chamberlin–Moulton theory effectively ceased after 1915 once the true nature of spiral nebulae, as whole galaxies, was understood.

Chapter 9

British Big Tides

There is a tide in the affairs of men which, taken at the flood, leads on to fortune.

William Shakespeare (1564–1616), *Julius Caesar*

9.1. The Jeans Tidal Theory

Although by 1915 the Chamberlin–Moulton model had been abandoned, the idea that a tidal interaction had taken place between two stars was taken further by the British astrophysicist James Jeans (1877–1946).

Jeans

The Jeans model, put forward in 1916, was quite different in that solar prominences were not involved. The basic idea was that a massive star passed close to the Sun and raised huge tides on it, so large that material left the Sun in the form of a filament. This filament then broke up into a series of blobs and each blob eventually collapsed to form a planet. The attraction of the retreating massive star on the blobs pulled them into orbits around the Sun. This interaction is illustrated in Figure 9.1.

The blobs, potential planets referred to as *protoplanets*, went into elongated orbits around the Sun. When each blob most closely approached the Sun, a smaller-scale version of the same process occurred in which the Sun pulled out a tidal filament from the protoplanet thus giving rise to satellite families.

What made this model different from those that preceded it was that Jeans was a good theoretician and he produced mathematically-based analyses of all the different aspects of it. He showed that under the tidal force of the approaching massive star the Sun would have become distorted into an egg shape and the profile at the small end of the egg would become sharper and sharper as the star approached more closely. Eventually the profile would give rise to a sharp point

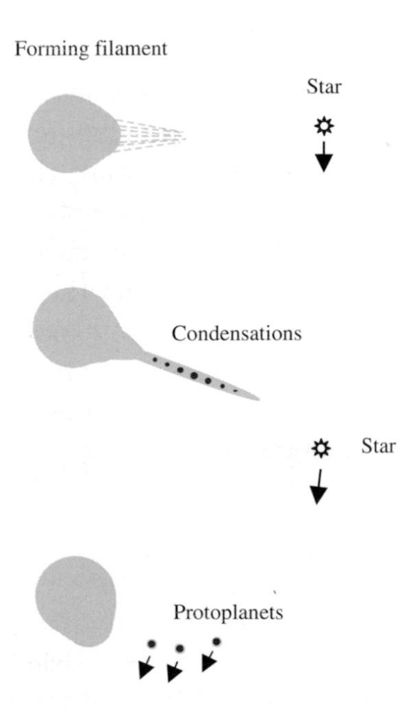

Figure 9.1. The Jeans tidal theory.

and thereafter material would escape from it in the form of a stream, or filament. Jeans then produced more analysis to show that the filament would break up into a series of blobs. The process is rather similar to something that is seen when a fine stream of water coming from a tap suddenly breaks up into a stream of droplets. For the water stream it is a property of a liquid called surface tension that causes the break up. For the gas filament it is the force of gravity that creates the blobs. If the material in the gas stream has temperature T (absolute scale) and density ρ then the length of each blob within the stream is given by

$$l = \sqrt{\frac{\pi k T}{G \rho \mu}} \tag{9.1}$$

where k is *Boltzmann's constant* 1.38×10^{-23} J kg^{-1} K^{-1} (joules per kilogram per degree absolute), G is the *gravitational constant*, 6.67×10^{-11} m^3 kg^{-1} s^{-2} (metres cubed per kilogram per second squared) and μ is the mean mass of the molecules that constitute the gas (in kilograms).

Another important theoretical concept introduced by Jeans is the idea of a *Jeans critical mass*. If one imagines a uniform sphere of gas with a particular density and temperature then the Jeans critical mass defines the minimum mass for which the sphere would just be able to collapse. The gas sphere is under two influences — gravity that is tending to cause it to collapse and thermal energy due to its temperature that will tend to cause it to expand. If the mass is too small then the thermal forces are dominant; if the mass is larger than the critical mass then the gravity force dominates and the sphere can collapse. The critical mass for a sphere can be derived theoretically as

$$M_J = \sqrt{\frac{375 k^3 T^3}{4 \pi G^3 \mu^3 \rho}}. \tag{9.2}$$

For an irregular shape the numerical constant in (9.2) would be different but the dependence on the properties of the gas would be as indicated. If the mass of a blob in the tidal filament exceeded the appropriate Jeans critical mass for their shape then it would be able to collapse to form a planet.

This theory really captured the imagination of the astronomical community and it had very wide acceptance. It was a dualistic theory so it did not have to explain the slow spin of the Sun. It explained the formation of planets, all in direct orbits, and satellites for those planets. Jeans also pointed out that the filament would have a cigar shape and be thickest in the middle when the massive star was closest to the Sun so pulling out material at a greater rate. This would explain why Jupiter and Saturn, the most massive planets, were in the centre of the system. There was no direct explanation of the 7° tilt of the solar spin axis but the spin axis would certainly not be expected to be perpendicular to the plane of the planetary orbits. Those orbits would be close to the plane defined by the motion of the massive star with respect to the Sun. Although there were some loose ends to be tidied up it seemed that, at last, a viable theory for the origin of the Solar System had been produced.

9.2. Jeffreys' Objections

The British geophysicist Harold Jeffreys (1891–1989) was at first a strong supporter of the Jeans model, so much so that it was sometimes referred to as the Jeans–Jeffreys theory although the two individuals never worked together. Nevertheless, in 1929 Jeffreys was the first to express doubts about the idea. The first objection he raised was that the probability of a massive star passing close to the Sun was extremely small. That was not much of an argument. There is a tenet, known as the *anthropic principle*, which is generally accepted, that states that astronomical theories are constrained by the necessity of allowing human existence. That humankind exists cannot be disputed. If it is absolutely necessary for some highly improbable event to have taken place for we humans to exist then that event *did* take place.

The second of Jeffreys' arguments was based on some fairly subtle physics involving a mathematical quantity called *circulation*. What it amounted to was the statement that if the material for Jupiter

Jeffreys

was drawn out of the Sun, then because Jupiter and the Sun have virtually the same density, they should be spinning at the same rate. In fact, Jupiter spins about sixty times as fast as the Sun. Once again this turns out not to be a strong argument since it assumes that the Jupiter material came out of the Sun with the density it has now. If it came out at a lower density and then contracted, its faster spin can be explained.

In order to deal with this second argument Jeffreys suggested that instead of a tidal interaction, the star physically sideswiped the Sun so imparting spin to the material that was knocked off to form planets (Figure 9.2). There was some inconsistency here as physical contact would be much more unlikely than the tidal interaction that Jeffreys had criticized on the basis of its small probability. In fact, Jeffreys had returned to a model similar to that proposed in 1745 by the French astronomer Georges compte de Buffon (1707–1788). He proposed that a comet had grazed the Sun knocking off material that travelled out to various distances to form the planets. Buffon had no idea of the nature of comets and we know now that their masses are far too small to have much effect on the Sun.

Jeffreys' idea of a collision between a star and the Sun did enjoy considerable support and for some time was preferred to the Jeans model.

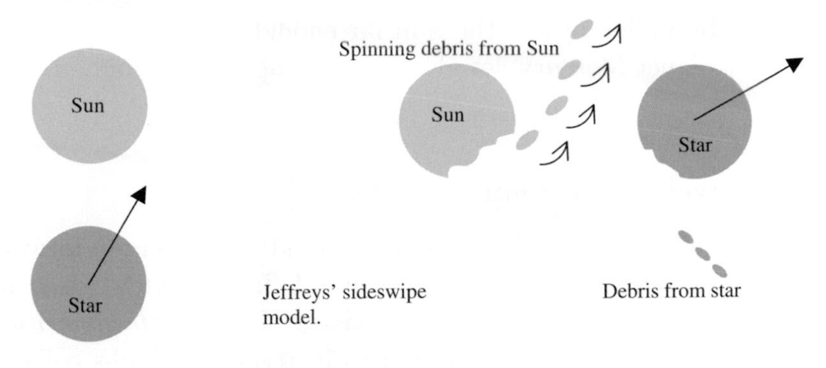

Figure 9.2. Jeffreys' star-collision theory.

9.3. Russell's Objection

The next objection raised against the Jeans theory, and Jeffreys' modification of it, was made in 1935 by the American astronomer Henry Norris Russell (1877–1957). The elliptical orbit of one body around another is closed, meaning that it follows the same path in space. Hence, if material is pulled out of the Sun into an elliptical orbit then its path will take it back to the Sun and it will be reabsorbed. Actually the Jeans model was a little more complicated than this because a blob that formed in the filament was affected by the gravitational attractions of the other blobs and of the massive star. Even with these influences Russell was able to show that material

Russell

could not be pulled out of the Sun far enough to explain even the innermost planet Mercury, let alone those much further out.

9.4. Spitzer's Objection

The American astrophysicist Lyman Spitzer (1914–1997) put forward an argument against both the Jeans and Jeffreys models based on theory produced by Jeans himself! This involves the *Jeans critical mass* that has already been referred to in respect of blobs collapsing to form planets. Spitzer argued that if the Sun had been in its present condition then an amount of material taken from it sufficient to produce Jupiter would have about the present density of Jupiter and a temperature of one million degrees Kelvin. The thermal force, tending to give expansion, would overwhelm the gravity force and the mass of gas would not stay together to form Jupiter but would violently dissipate.

Spitzer

In summary Russell's and Spitzer's arguments amounted to saying that not only could planets not form at sufficient distance from the Sun but they could not form at all!

9.5. A Modern Objection

Jeans accepted the validity of the objections made against his theory and wrote, in a truly objective spirit, "The theory is beset with difficulties and in some respects appears to be definitely unsatisfactory". At the end of this episode one had a model that could not be upheld but one also had a body of theoretical analysis developed by Jeans that was sound and could perhaps be deployed in theories yet to come.

Much later, long after the Jeans model had been abandoned, an objection was raised that would apply to both the Jeans model and to Jeffreys' modification of it. The temperature inside the Sun is high enough for many nuclear reactions to take place that involve atoms of low mass. Amongst these 'light atoms' are lithium, beryllium and boron that are transformed into other atoms at solar temperatures. By looking at the Sun's spectrum we can tell what sort of atoms are present and these light elements are conspicuously absent. However, on Earth these elements, whilst not abundant, are present in substantial quantities. The implication is that the material of the Earth did not come from a source as hot as the Sun — and by implication this would also apply to the other planets. This objection indicates that we must only consider theories for which planetary material has a 'cold' origin where the term 'cold' in this context can include temperatures up to hundreds of thousands of degrees, but no greater.

Chapter 10

Russian Cloud Capture — With British Help

Stooping through a fleecy cloud.

John Milton (1608–1674)

10.1. The Schmidt Model

The demise of the two dualistic theories, that had been dominant for the first three or four decades of the twentieth century, did not see an end to such theories. The next form of dualistic theory, the Accretion Theory, of a new and interesting kind, came from a Russian planetary scientist, Otto Schmidt (1891–1956) in 1944. Observations with telescopes show regions of the sky where stars cannot be seen because the light coming from them is absorbed by dense cool clouds of gas and dust. Schmidt argued that from time-to-time a star on its journey through the galaxy would pass through one of these clouds and, in doing so, might pick up a surrounding envelope of gas and dust. The envelope would settle down into a disk-like form and from its material the planets would form.

Schmidt assumed that if only the Sun and the cloud were present then capture of cloud material could not take place. This assumption was based on the principle that if two bodies approach each other from an infinite distance and do not collide then they must end up an infinite distance apart — which is certainly true for two point masses but hardly applies to a star and a very extended cloud. For this mistaken reason Schmidt postulated that there had to be another star somewhere in the vicinity, the gravitational field of which would

Schmidt Lyttleton

enable the capture to occur by removing some energy from the interacting star-cloud system. This made the likelihood of the event very small although, appealing once again to the anthropic principle, it did not necessarily rule out the idea. In 1960, the British astronomer Ray Lyttleton (1911–1995) took up Schmidt's idea and showed that it was not necessary to have another star present and that the Sun, after passing through the cloud, could be accompanied by captured material (Figure 10.1).

10.2. Lyttleton's Modification of the Accretion Theory

Lyttleton used theory developed by Bondi and Hoyle that dealt with how a star could capture material from the cloud. Figure 10.2 shows the motion of the cloud material, at speed V relative to the star. The cloud material is attracted towards the axis by the gravitational

Figure 10.1. The Sun picking up a dusty gaseous envelope after passing through a dense cool cloud.

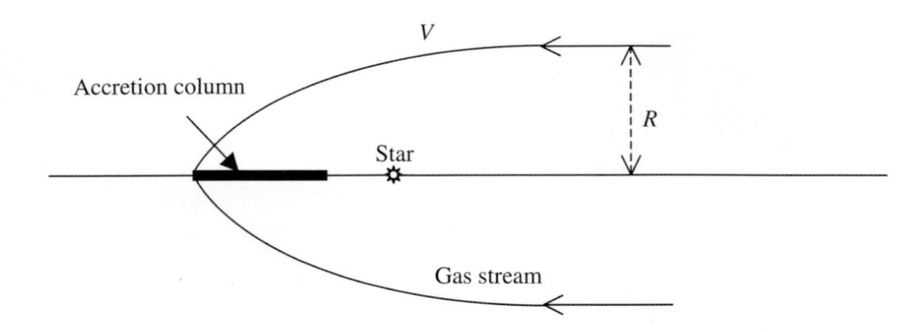

Figure 10.2. Gas streams interacting on the axis.

attraction of the star; the figure shows streams coming in from distance R on opposite sides of the axis.

When the gas streams interact on the axis they mutually destroy their components of motion perpendicular to the axis, which slows the material down to less than the escape speed from the star. This can happen for streams coming from up to some maximum distance from the axis and the effect is to create a high density accretion column of material on the axis that is captured by the star. Some of it may impinge on the star, but if there is some residual tangential motion of the colliding streams, then this will lead to capture as a surrounding envelope of material.

The original Lyttleton model was unsatisfactory in that it postulated unrealistic conditions — for example it assumed that the Sun would enter the cloud with a speed of 0.2 km s^{-1}, which is impossibly low since the gravitational attraction of the cloud on the Sun would alone have given a greater relative speed than that. However, in 1973, Chris Aust and I showed that some modifications of the original conditions specified by Lyttleton could lead to the Sun being surrounded with about the right total mass of material at about the right distances to produce the planets. What was missing was some mechanism for turning this very diffuse cloud material into planets — a problem that occurs with other more modern theories that will be described later.

In view of its rather vague nature, the theory did not enjoy much support — but the idea of capture was to return.

Chapter 11

German Vortices — With a Little French Help

"Mind and matter," said the lady in the wig, "glide swift into the vortex of immensity".

Charles Dickens (1812–1870), *Martin Chuzzlewit*

11.1. First Ideas about Vortices

When a spoon is moved through a cup of tea or coffee the liquid surface moves around in a set of swirling motions. Larger-scale motions, described as *vortices*, are usually accompanied by smaller swirling motions, known as *eddies*. Such motions are also seen when a flowing fluid moves past a stationary object, such as a post set in a river. We have already referred, in relationship to Laplace's nebula theory, to the work of René Descartes, the French philosopher, mathematician and physicist, who was a contemporary of Galileo and Kepler and who, like them, believed in the essential correctness of the Copernican model. Descartes' theory dealt with the heliocentric nature of the Solar System, the planarity of the system and the direct orbits of the planets and satellites. Descartes described this idea in a book *le Monde* but, mindful of the way that the Church had treated Galileo he was fearful to publish it and so it did not appear until 1664, after his death.

Von Weizsäcker

11.2. The von Weizsäcker Vortex Theory

In 1944, the German astrophysicist Carl von Weizsäcker (1912–2007) considered how a pattern of vortices might be set up in gaseous disk due to turbulence, random motions of a fluid that stir it up in the way that a spoon stirs up liquid in a cup. He suggested that a pattern of vortices was set up as shown in Figure 11.1.

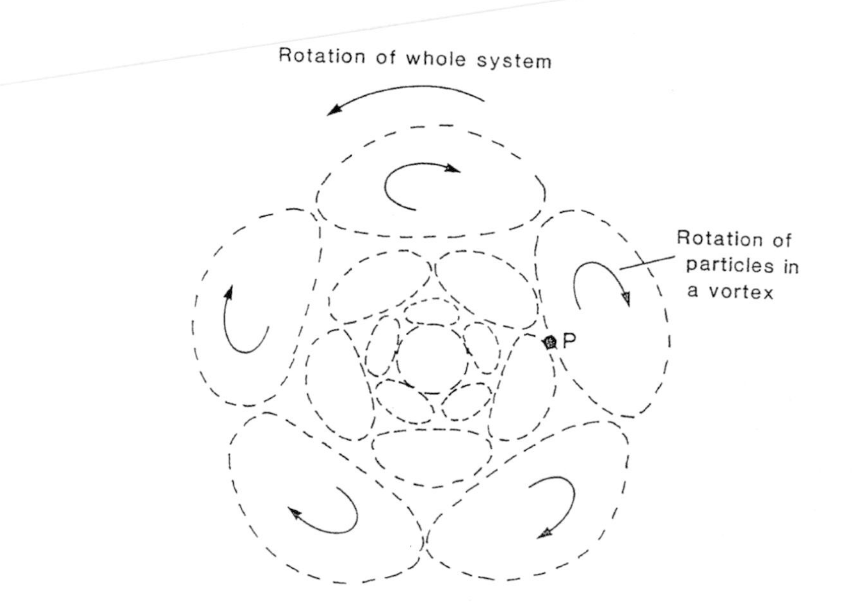

Figure 11.1. The pattern of vortices according to the von Weizsäcker model.

He showed that the combination of a clockwise rotation within each vortex with an anticlockwise rotation of the whole system could lead to each individual particle of fluid moving in an elliptical orbit around the central mass, just as Kepler's first law requires. At a point such as P, where neighbouring vortices met, material would be colliding at high speed and material would coalesce there to form condensations. All the condensations in one ring would eventually combine to form a planet. Von Weizsäcker showed that if there were five vortices to a ring then the pattern of orbital radii would be similar to that observed in the Solar System. As with all evolving disk models the central part of the disk was assumed to eventually form the Sun.

11.3. Objections to the Vortex Idea

The harshest critic of this model was Harold Jeffreys. Von Weizsäcker had appealed to turbulence as the agency for enabling the pattern of vortices to be established. Turbulence causes a loss of energy of a system since colliding turbulent streams of matter generate heat and this heat is then radiated away. The heat energy is derived from the mechanical energy and hence the mechanical energy of a turbulent system gets less. Jeffreys showed that von Weizsäcker's system of vortices was a *high* energy system and hence would not form as a result of turbulence. The natural end state for a turbulent system is one of lowest energy and for a disk of the type postulated by von Weizsäcker, this would lead to all parts of it moving in smooth circular motion around the central mass.

The proposed mechanism deals neither with the problem of the formation of a slowly spinning Sun nor with the formation of satellites, both of which are basic requirements for any plausible theory.

Chapter 12

McCrea's Floccules

...he made the stars also.

Genesis

12.1. Producing Stars and Planets Together

The general idea behind two of the previously described theories is that an individual mass of material undergoes processes that produce the Sun accompanied by the planets. Observations show that many stars occur in clusters and indeed it is generally thought that all sun-like stars originally formed in clusters. This will be discussed further in Chapter 26. In 1960, William McCrea (1904–1999) took a rather broader view than hitherto and considered a scenario that would produce a whole cluster of stars together with planets for some of them. He tried simultaneously to solve the problem of why it is that solar-type stars spin slowly and how it is that they acquire planetary systems.

McCrea

McCrea's starting point was a cloud of gas, mostly hydrogen and helium, containing about 1% of its mass in the form of dust. This cloud, which will eventually form a cluster, is in a turbulent state so that individual regions of the cloud are moving in random directions with supersonic speeds. When two of these regions collide, the gas in them is compressed and McCrea modelled the cloud as a large number of high-density regions, called *floccules*, moving in random fashion in a lower-density background. In the first version of the theory the masses of the floccules were about three times the mass of the Earth. When floccules collided they combined and the larger-mass body so formed is more able to attract more floccules thus enhancing its growth further. In this way, in each region, one dominant body formed a *protostar* that was destined eventually to become a star. Smaller clusters of floccules became incipient planets, *protoplanets*, that were captured in orbit around the protostars. A schematic representation of this model is shown in Figure 12.1.

A good feature of this model is that it leads to a slowly spinning Sun. Every time a floccule joined the growing protostar it contributed to its spin according to the speed of the floccule and the angle of its impact; the floccule marked with a cross in the two-dimensional figure contributed to the spin in a clockwise direction. In the three-dimensional situation the contributed spins would have been in all

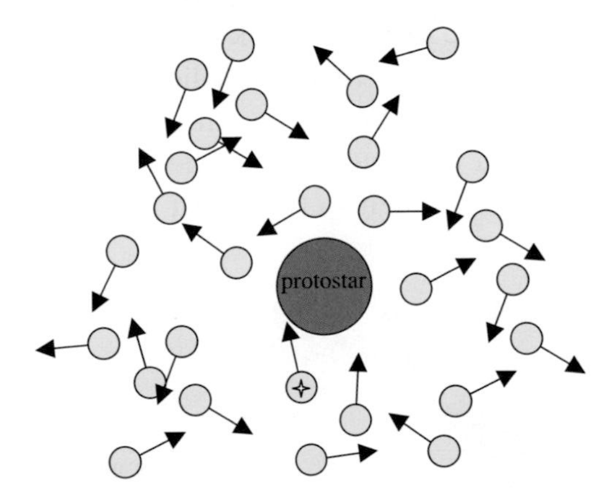

Figure 12.1. Two-dimensional representation of floccules and a growing protostar.

directions at random and they would have tended to cancel out. McCrea showed mathematically that the resultant spin of the Sun would have been very close to the expected value.

Another prediction of the model is that the initial protoplanets would have been more massive than the present planets and that they would have spun very rapidly. This is because it took fewer floccules to form a planet so the cancelling-out process for the spins would have been less efficient. This apparent difficulty with the model turned out to be a positive feature. As the protoplanet collapsed, it would have spun faster and faster until eventually it became unstable and broke up. The form of break up in such a situation had already been investigated by Lyttleton in considering another theory. The protoplanet would break up into two parts with one part having about eight times the mass of the other. As the two parts separated so a fine neck, or filament of material, connected them and little blobs forming in this (as Jeans described) would be retained in orbit by the larger part (Figure 12.2). These constituted a satellite family. The smaller portion would have moved at a much greater speed than the larger one and its speed would have been so great that it would have been able to leave the Solar System. One is left with a slower-spinning protoplanet and a satellite family, just what is required for the major planets.

McCrea suggested that in the inner part of the Solar System, the process of disruption into two parts took place involving just the dust component of the material so that two solid terrestrial-type planets were formed. He suggested Venus-Mercury and Earth-Mars as the two pairs produced. In the inner part of the Solar System the speed required to escape is very high so that both parts were retained.

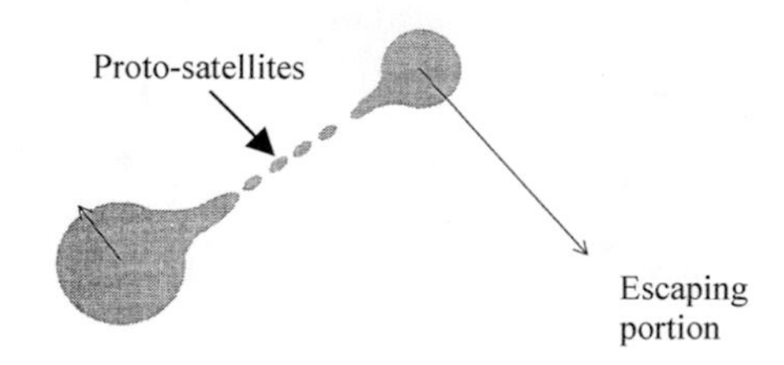

Figure 12.2. The break-up of a rapidly spinning protoplanet.

By assuming that all the floccules in a particular region of the whole cloud were bound to just the nearest protostar, McCrea showed that the number of planets and their orbital radii would be in reasonable agreement with the present Solar System. The theory was a bit lacking in detail and was not computationally modelled but the numerical results were very satisfactory.

12.2. Objections to the Floccule Theory

The first objections to the floccule theory were raised by myself in a private communication to McCrea. The theory developed by McCrea gave the density and temperature of the floccules and it turned out that the proposed floccule masses of three times that of the Earth were far less than the Jeans critical mass. Thus the floccules would have expanded and would have completely dissipated on a timescale of about a year. This would not matter if several floccules could get together to make a larger mass in a short time but, again using McCrea's own figures, the average time for two floccules to get together is about thirty years. McCrea countered this objection by increasing the masses of the floccules to about the mass of Saturn, which solved the floccule instability problem but introduced others. For example, with fewer floccules forming the Sun the process by which the spins induced by individual floccules cancelled out would be less efficient and the final spin rate of the Sun would have been much higher. There is also the difficulty that the whole basis of producing floccules of planetary mass is suspect. In Chapter 26 an analysis is described that indicates that the first condensations produced in a cloud are of stellar rather than of planetary mass.

Another problem was McCrea's assumption that each floccule was bound to the nearest protostar and to no other. With this assumption he was able to show that the total angular momentum of the floccules captured by a protostar would be similar to that of the solar-system planets. In fact the individual floccules would have travelled throughout the cloud, through regions dominated by different protostars, and a great deal of their energy would have gone into producing relative motion of the protostars rather than the motion of protoplanets around protostars.

Finally, we should note that the protoplanets could have been captured in orbits with any orientation so that an initially planar system of planets would not be expected and planets could be orbiting in either the direct or retrograde senses. There may have been some mechanism that pulled all the planets into one plane — for example, if the protostar was surrounded by a disk. Then, if there were fewer retrograde planets, they could have been removed by collisions with some that were in direct orbits leaving the surviving protoplanets in direct orbits to form the present system.

Chapter 13

What Earlier Theories Indicate

If men could learn from history, what lessons it might teach us!

Samuel Taylor Coleridge (1772–1834)

13.1. Angular Momentum Difficulties

Angular momentum problems occur in two different aspects in the various theories that have been described so far. The main repositories of angular momentum in the Solar System are the planets in their orbits — accounting for all but a tiny part of the total angular momentum — and the Sun's spin. Other sources, such as the spin of the planets and the orbits and spins of the satellites, are so tiny by comparison that they can be ignored.

The monistic theories that start with a single mass of material that is to form the Sun and the planets, have the great difficulty that angular momentum has somehow to be removed from the bulk of the material that will form the Sun and be transferred to the small amount of material that will become planets. Neither the Laplace nebula theory nor the von Weizsäcker vortex theory addressed the issue of a slowly rotating Sun. We shall see later (Chapter 18) that the need to solve this problem has led to new ideas about transferring angular momentum in a modern nebula model.

For dualistic theories the problem of the slow spin of the Sun is avoided — but it could be argued that a complete description of the origin of the Solar System should include an explanation for this important property of its dominant member. However, there are those that take the view that observations show that most Sun-like stars spin slowly and that this is a legitimate starting point for a theory. If this is accepted then the emphasis is just to explain the presence of the planets. But even with this let-out the angular momentum problem is not trivially resolved. An important problem with the Jeans model, and also with Jeffreys modification, is that material could not be removed sufficiently far from the Sun, either by the tidal action of a massive star or by a stellar collision. This is an angular momentum problem of a different kind where not enough of it is possessed by the planets. The dualistic Schmidt-Lyttleton theory of cloud capture and McCrea's floccule theory both get closer to solving the angular momentum problem. For cloud-capture the captured material can possess the right amount of mass and angular momentum to explain the planets. The situation with respect to the floccule model is less clear-cut since some of the underlying assumptions made by McCrea are suspect and the mechanism has never been properly tested by any kind of modelling procedure.

13.2. Planet Formation

While having material at the right distance from the Sun is a *necessary* condition for a plausible theory, that by itself is not *sufficient*. It must also be shown that the material forms planets.

None of the monistic theories we have described so far has even considered this problem in any detail. Laplace suggested that clumps in his rings would form by gravitational attraction and that then the clumps would combine. Actually it is possible to show that unless his rings had masses very much greater that that of planets, the rings would have been very unstable and would have dispersed to give a disk without rings in a very short time — much shorter than the time required for clumping to take place. The end result would be a fairly structureless disk within which the planets must form — a similar situation to that obtained with the cloud-capture model.

The floccule theory produces planets by concentrating cloud material through collisions. It is certainly true that colliding material would be compressed but it simply would not produce planetary masses in a large cloud. The turbulent streams in such a cloud would have had masses similar to the Jeans critical mass for the cloud material and these would have been of stellar mass. When they collided, stellar-mass condensations would have been produced.

The only early theory that hints at how planets could be successfully formed is that of Jeans. The break up of a filament into a set of blobs under gravitational effects is well founded theoretically and, as will be shown in Chapter 28, has also been successfully modelled. The problem with the Jeans theory was not that the mechanism for producing planets was unsatisfactory but rather that it was being applied to the wrong material. It is quite possible to have material in a filament at a density and temperature that would give planetary mass blobs with greater than the Jeans critical mass. This certainly requires that the material should be at a temperature much lower than that of typical solar material — but that requirement is also indicated by the quantities of the light elements lithium, beryllium and boron in the Earth's crust.

13.3. Indications of Requirements for a Successful Theory

The older theories do not clearly indicate the exact nature of a successful theory but they do offer some possible alternatives. The best indications are for a dualistic theory in which the formation of the Sun is not directly linked to the formation of the planets. However, it really should be necessary to explain the formation of the Sun, and by implication other stars, to have an entirely satisfactory theory of the origin of the Solar System. A dualistic approach has the advantage that it reduces the problem of the formation of the Sun to that of producing a solar mass with solar angular momentum, without the additional complication of *simultaneously* having to produce a much smaller mass, the planets, with much greater angular momentum.

In planet formation cold material is absolutely essential. For this reason it is not possible to have planetary material derived from the existing Sun. This requirement is satisfied by a monistic theory in which both the Sun and the planets are to be formed from cold material. If the material has the density and temperature that enabled a planetary mass of it to satisfy the Jeans mass criterion, clearly then this would be an advantage. Otherwise a process by which the more diffuse material is able to concentrate itself to form planets is required.

The planetary material would either originally have, or need to acquire, the angular momentum required to explain the final planetary orbits. If the potential planetary material and the Sun approached each other in such a way that the material initially had the required angular momentum relative to the Sun, then this would be helpful. This general condition is satisfied, more-or-less, by the Schmidt–Lyttleton theory and the McCrea floccule theory.

Bearing these indications in mind, we shall be looking at some modern theories to see how well they are satisfied. Before doing so we shall learn about some recent observations that indicate that there are planetary systems other than the Solar System. Any plausible theory of the origin of the Solar System will also need to explain the formation of these other systems.

NEW KNOWLEDGE

Chapter 14

Disks Around New Stars

Look at the stars! Look, look up at the skies.

Gerard Manley Hopkins (1844–1889), *The Starlight Night*

14.1. How Hot and How Luminous?

An important property of any star is its temperature and there is an associated property, its *luminosity,* which is a measure of the rate at which it emits energy by radiation. The radiation it emits is *electromagnetic (e.m.) radiation*, a coordinated electrical and magnetic disturbance propagated in the form of a wave. The wavelength dictates what kind of radiation it is and how it may be detected. For example, if the radiation has a wavelength between 0.4 microns[1] and 0.7 microns then the radiation is visible light with blue and red at the short and long wavelength ends respectively. Going from blue towards shorter wavelengths first traverses the ultra-violet (UV) region, then X-rays and finally very short wavelength γ rays with wavelengths from one ten thousandth to one ten millionth of that of visible light. Starting from the red end and going towards longer wavelengths we first pass through the infrared region which gradually merges into radio waves with wavelengths up to one hundred thousand million times the wavelengths of visible light. This range of electromagnetic radiation is illustrated in Figure 14.1; it is interesting to note how little of it we can detect with our eyes!

[1] one micron (μ m) = one millionth of a metre.

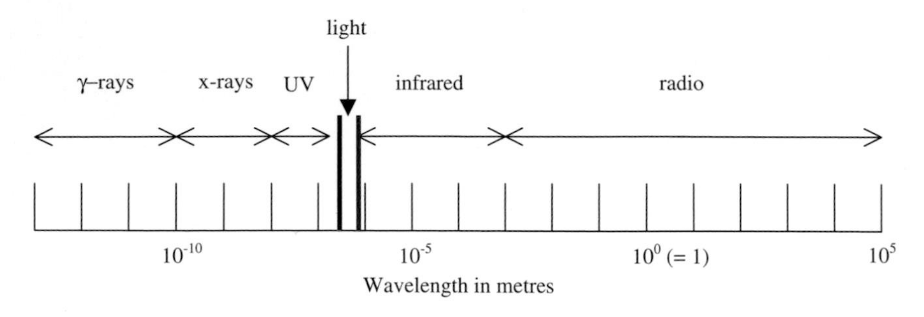

Figure 14.1. The electromagnetic spectrum.

It is a matter of everyday experience that when bodies are heated to a high enough temperature they emit light and the colour of the light changes with temperature. A piece of iron heated to a temperature too low to emit light (i.e. an iron for pressing clothes) will still emit radiation in the form of heat, which we can think of as in the infrared region. Heated a little more the iron gives visible light that is at first a very dull red but becomes a brighter red as the temperature increases. As the temperature increases further, the light from the iron changes first to a yellow colour and then becomes a brilliant white. At an even higher temperature the white light becomes tinged with blue. As it is for a piece of iron, so it is for a star. Lower temperature stars shine with a reddish light while high temperature stars look blue. Hence from the colour of a star we can determine its temperature.

At any temperature, a hot body will emit radiation over a wide range of wavelengths and the overall visual effect, if there is one, depends on the relative amounts of the various parts of the spectrum. A body at the temperature of the Sun has its main visible emission in the green to red part of the spectrum — which makes it appear as a yellow star. A very hot star will have peak emission in the ultraviolet region. Over the visible part of the spectrum there is high emission everywhere but more at the blue end than the red end — hence the appearance of bluish-white. A comparatively cool object, say at the temperature of boiling water, will emit infrared radiation that can be detected as heat but is invisible to the eye.

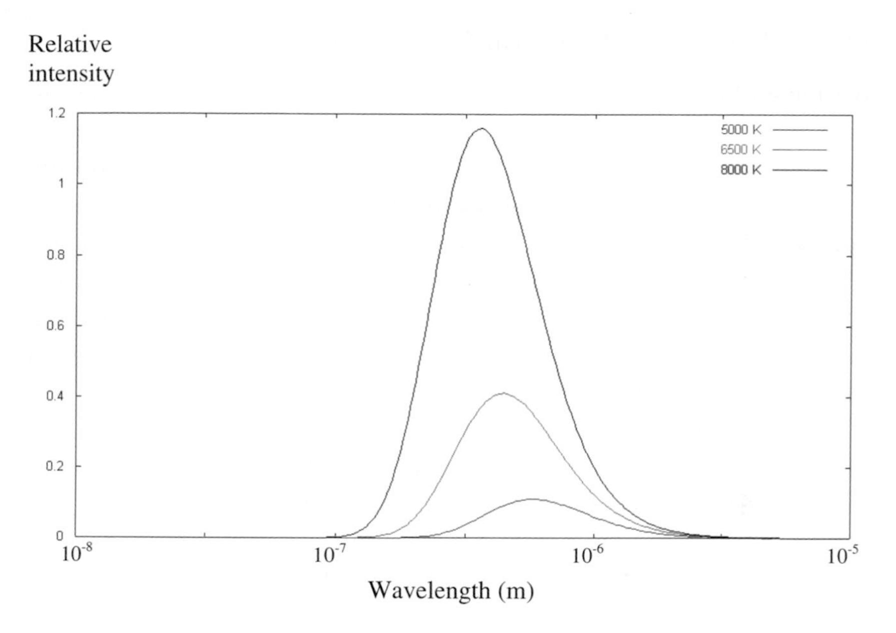

Figure 14.2. Radiation curves at different temperatures.

In Figure 14.2, the relative intensity of radiation from a heated object for different wavelengths is shown for the temperatures 5,000 K, 6,500 K and 8,000 K. First it will be seen that the peak of the curve moves towards shorter wavelengths for higher temperatures. The other very noticeable feature is that the total intensity from the complete range of wavelengths increases very rapidly with temperature; theory shows that it is proportional to the fourth power of the temperature so it is more than 6 times greater at 8,000 K than it is at 5,000 K. The total amount of light coming from an object at a particular temperature is proportional to the area of the radiating surface so the luminosity of a star thus depends on the fourth power of its temperature multiplied by its surface area.

The brightness of a star as seen from Earth will depend both on its luminosity and its distance. A car headlight that is dazzling when seen close at hand seems much less bright when the car is at a great distance. There are various ways of estimating the distance of stars and if the brightness of a star is measured and its distance is known, then its luminosity can be found.

14.2. What is a *New* Star?

We have already used the term *protostar* to indicate an object that will eventually become a normal star. When a protostar is formed as an identifiable entity, it is a large nebulous ball of cool gas and dust that slowly at first, and then ever faster, begins to collapse. Its initial radius will be something in the range 1,000 to 2,000 au or from 30 to 60 times the distance of Neptune from the Sun. At first the protostar is so transparent that the heat generated within it by the collapse can readily be radiated away but as it becomes denser, it becomes more opaque and the heat within it is retained. Eventually it stops collapsing quickly and becomes a hot body at a temperature of a few thousand degrees. It is radiating energy and, as it does so it slowly collapses and becomes hotter. It may seem odd that a radiating body becomes hotter but the extra energy required both to heat it up and to radiate away comes from its collapse. We may refer to a star in this state as a *new* star.

When we refer to the temperature of a star we usually mean the temperature of the surface that we can see, so that the temperature of the Sun is about 5,800 K. However, the temperature is much higher inside the star and in the case of the Sun the maximum temperature within it is about 15 million K. So it is for a newly forming and evolving star. The *new* star in its slowly evolving state has an external temperature of, say, 4,000 K but it may be several million K internally and will steadily rise as the external temperature rises. When the internal temperature becomes high enough a critical point is reached when nuclear reactions can occur, involving the transformation of hydrogen into helium. Once this source of energy is available the star is said to be on the *main sequence* and thereafter its temperature and luminosity will vary comparatively little over a considerable period of time. The Sun is a main sequence star. It has been on the main sequence for about 5,000 million years and will remain on it for another 5,000 million years.

Because the temperature and luminosity hardly change for a star on the main sequence it is impossible to precisely judge its age. We can do so for the Sun by assuming that it has the age of the Solar

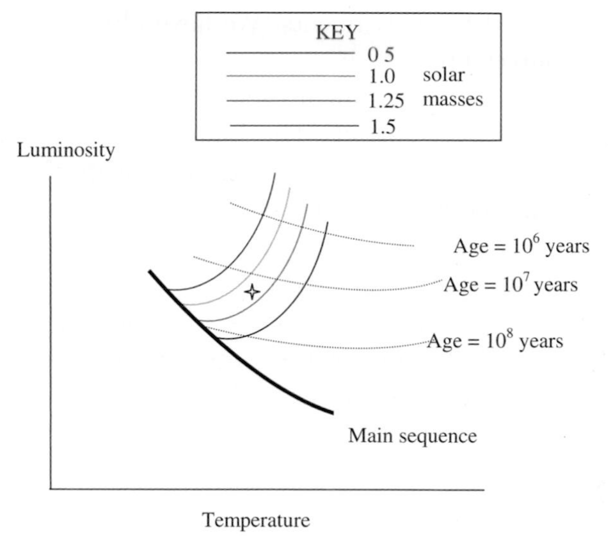

Figure 14.3. Pathways towards the main sequence.

System that can be dated in various ways not involving the Sun itself. But when the *new* star is on the slowly evolving path *towards* the main sequence both its temperature and luminosity *are* changing and can give an estimate of its age. In Figure 14.3 there is shown a schematic representation of the temperature and luminosity curves for protostars of various masses in their final journeys to the main sequence as found from theory. If the temperature and luminosity of a *new* star is measured then it is possible to determine both its mass and its age. For the observation represented by the four-pointed star the *new* star has a mass of about 1.1 solar masses and an age of approximately thirty million years.

14.3. Detecting Disks

A *new* star will be recognized by its temperature and luminosity but what can also be measured is the distribution of energy with wavelength, as shown in Figure 14.2, which is a well recognized curve for any particular temperature. In the mid-1980s, it was found that for a proportion of *new* stars the curve seems to have an extra bump in the infrared part of the spectrum, which is outside the visible range but can be detected by instruments. We have already seen that low temperature sources are very inefficient radiators in terms of rate of energy output per unit area but if they have a large area then they can give a considerable output. The bumps on the curves of *new* stars are called an *infrared excess* and they indicate the presence of a low-temperature, large-area radiating body accompanying the star. This is, quite reasonably, interpreted as being due to a large dusty gaseous disk encompassing the star.

From observations of many *new* stars it has been determined that the disks have a finite lifetime as there are very few disks for stars of more than about three million years old. Lifetimes from one to six million years are inferred, with most being at the lower end of the range. Other kinds of observations suggest what the disks are like. They seem to have masses mostly in the range one-hundredth to one-tenth of a solar mass with radii anything from tens to hundreds of astronomical units.

Various ideas concerning planetary formation involve the presence of disks so confirmation of their existence and the determination of their properties is of great interest.

Chapter 15

Planets Around Other Stars

Observe how system into system runs, what other planets circle other Suns.

Alexander Pope (1688–1744)

15.1. Stars in Orbit

It is customary to refer to a planet being in orbit around the Sun, which is a correct statement in itself, but the mental picture that this evokes is of a planet moving around a *stationary* Sun. The Solar System is complicated, with many planets, so to simplify the discussion we shall consider a star with a single planet. How should we describe the actual motion for this simple system?

What is really happening is that the two bodies, the star and the planet, are *both* in orbit. Their motion is such that they both go round a fixed point, the *centre of mass*, which lies on the line connecting them (Figure 15.1). The distances of each from the centre of mass is proportional to the mass of the *other* body; this means that if the star is 500 times as massive as the planet then the planet is 500 times more distant from the centre of mass.

The stellar orbit as seen in the figure is greatly exaggerated in scale for clarity of presentation. If the Solar System consisted just of the Sun and Jupiter, then since the Sun is 1,000 times as massive as Jupiter, the orbit of the Sun would be 1,000 times smaller, and the orbital speed of the Sun 1,000 times smaller than that of Jupiter. The speed of Jupiter in its orbit is 12 km s^{-1}, which means that the

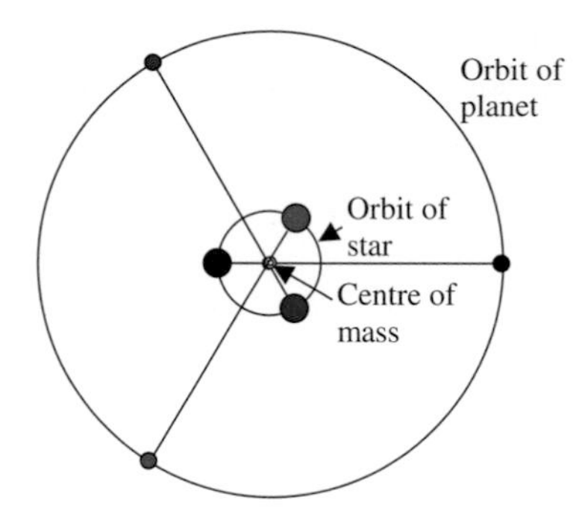

Figure 15.1. The orbits of a star and its planet.

speed of the Sun in its orbit is 12 m s^{-1}. A similar situation would occur for any star with an accompanying planet.

These relationships can be expressed as

$$\frac{r_S}{r_p} = \frac{v_S}{v_P} = \frac{M_P}{M_S} \tag{15.1}$$

where r, v and M indicate distance from the centre of mass, orbital speed and mass respectively and subscripts S and P indicate the star and the planet.

It would be extremely difficult, although perhaps not impossible, to visually detect a planet in the immediate vicinity of a star since the light of the star would overwhelm the light reflected off the planet. However, if we imagine that the line of sight from the Earth was in the plane of the star's orbit, then as the star orbited, the star would be seen first retreating from and then approaching the Earth in a periodic fashion. If this motion could be detected and measured, then although we could not see the planet directly we could see a consequence of its presence and thereby be able to learn something about it.

15.2. Finding the Speed of a Star

Imagine being present at a Formula I Grand Prix observing from the middle of a long straight lane. The cars approach with their finely tuned engines screaming at a high frequency corresponding to the speed with which the engines are revving. They then draw level with the observing position and thereafter they are retreating. But what has happened? The note from the cars has noticeably changed, with the frequency now much lower. Have the drivers all throttled back? No, they have not, and what has been heard is due to a well known phenomenon in physics — the Doppler-shift effect.

Sound is a wave motion consisting of variations of air pressure propagating with a particular wavelength and a particular pitch (frequency). The Doppler-shift is the process by which an approaching source has a higher perceived pitch and a retreating source a lower one. It can also be expressed in terms of wavelength where the wavelength of an approaching sound source is shortened and that of a retreating source is lengthened. In a conceptual way we may consider that the approach of a sound source compresses the sound waves and makes them shorter while the retreat of the source stretches them out as illustrated in Figure 15.2.

What is true for sound is also true for light, which is another type of energy propagated in the form of waves. The light from a source that is moving away from an observer is seen with a longer wavelength, i.e. redder, than if it is at rest. Conversely, if it is moving towards the observer then the light appears to be of shorter wavelength i.e. bluer. This then provides us with a way of detecting the motion of a star since a star is after all a source of light. However,

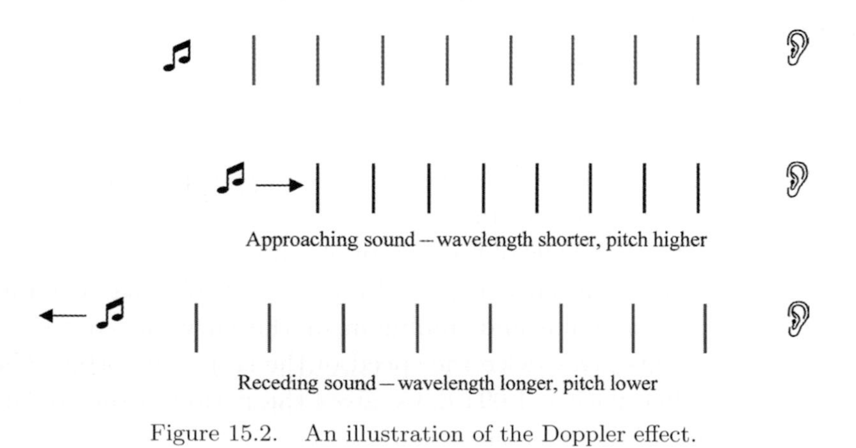

Figure 15.2. An illustration of the Doppler effect.

Figure 15.3. Absorption (Fraunhofer) lines in the solar spectrum.

we should not be able to detect the motion just by looking at the overall colour of the light since the change would be far too small to detect in that way. Fortunately the spectrum of a star, just like that for the Sun, contains a large number of dark lines (Figure 15.3) corresponding to light that has been subtracted from the spectrum due to absorption by various kinds of atoms. These act like markers for particular wavelengths of light and we can detect motion by the changes in wavelengths of these lines.

To determine the speed of the star, v_S, either towards or away from the Earth requires a measurement of the changes in the wavelengths of the lines. We can then use the Doppler-shift formula

$$\frac{v_S}{c} = \frac{\Delta\lambda}{\lambda} \tag{15.2}$$

in which c is the speed of light, λ the wavelength of a spectral line and $\Delta\lambda$ the change in the wavelength. The speed of light and the wavelength of the line are known so that a measurement of the change of wavelength gives the speed of the star. If the change of wavelength is positive then the star is moving away from the Earth, or if negative, towards the Earth.

Of course, there are a few complications in these measurements since the Earth itself is moving around the Sun and the centre of mass of the star-planet system may also be moving relative to the Sun. Nevertheless, it is possible to allow for these factors and explicitly to determine the speed of the star in its orbit.

Figure 15.4 shows some of the first observations of variable stellar velocities taken over a ten-year period for the star 47 Uma. The difference of the maximum and minimum of the curve fitted to the observation points gives twice the speed of the star in its orbit. The period of the fluctuation, 1,094 days, gives the period of the stellar, and hence also of the planetary orbit.

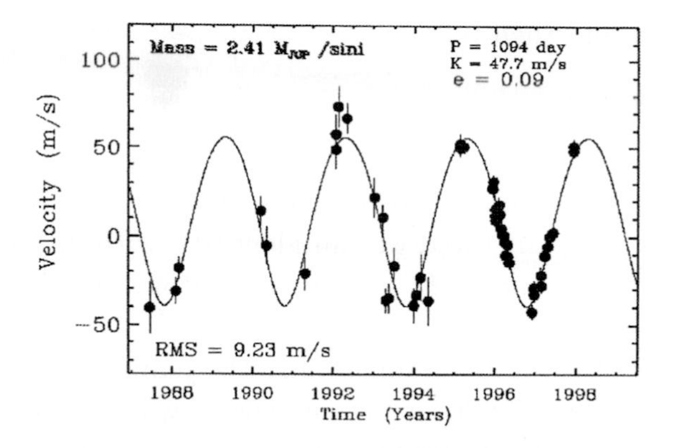

Figure 15.4. Stellar velocity measurements for 47 Uma.

15.3. Finding Out About the Planet

Knowledge of the speed of the star in its orbit and its orbital period is not sufficient information to enable an estimate to be made of the mass and orbit of the planet. Another necessary piece of information is the mass of the star and fortunately, we have a way of estimating this. In the previous chapter, we introduced the idea of a main sequence star, one that is converting hydrogen to helium and which is in a long-lasting state. For main sequence stars the temperature and mass are related so that if we know the temperature of a star then we can also estimate its mass. The stars that have been found to possess planets are all main sequence stars so by measuring their temperatures their masses can also be estimated. Actually, astronomers have a more subtle way of determining where the star is on the main sequence by looking at the pattern of the strength of the absorption lines in their spectra corresponding to different atoms, but describing the process in terms of temperature illustrates the general principle.

From the theory of planetary motions developed by Newton, if the mass of the star, M_S, and the period, P, of the orbit are known then so is the semi-major axis, a, of the orbit and the speed of the planet, v_P, in that orbit. This comes from the relationships

$$P = 2\pi\sqrt{\frac{a^3}{GM_S}}, \tag{15.3}$$

which gives a and then

$$v_P = \frac{2\pi a}{P} \tag{15.4}$$

that gives v_p. Equation (15.3) assumes that the star is much more massive than the planet and we can identify a with r_P for a circular orbit. With v_P, v_S and M_S known, then from (15.1) M_P can be found.

All the analysis above assumes that the plane of the star-planet orbit is in the line of sight, which in general will not be true. What the Doppler-shift measurements give is the speed along the line of sight that will normally be smaller than the actual speed of the star. The relationship between the speed along the line of sight and the actual speed of the star is shown in Figure 15.5.

Since the estimate of the mass of the planet is proportional to the speed of the star, if we underestimate the speed of the star then we also underestimate the mass of the planet. For this reason we are only able to estimate the *minimum* planetary mass and the true mass of the planet will be greater by an unknown factor.

There is a situation, by its nature rare, where an actual planetary mass can be estimated. This is when the line of sight is so close to the plane of the orbit that the planet actually transits the star — that is to say that the disc of the planet moves across the disk of the star. That this has happened is indicated by a diminution of the light as the planet makes its passage. An excellent example of such an observation is shown in Figure 15.6. The proportion of reduction in the brightness of the star gives the ratio of the cross sectional area of the planet to that of the star. Since the radii of main sequence

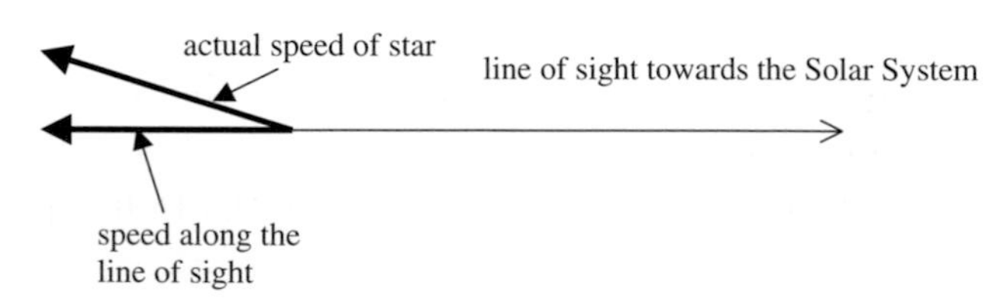

Figure 15.5. The speed along the line of sight is smaller than the speed of the star.

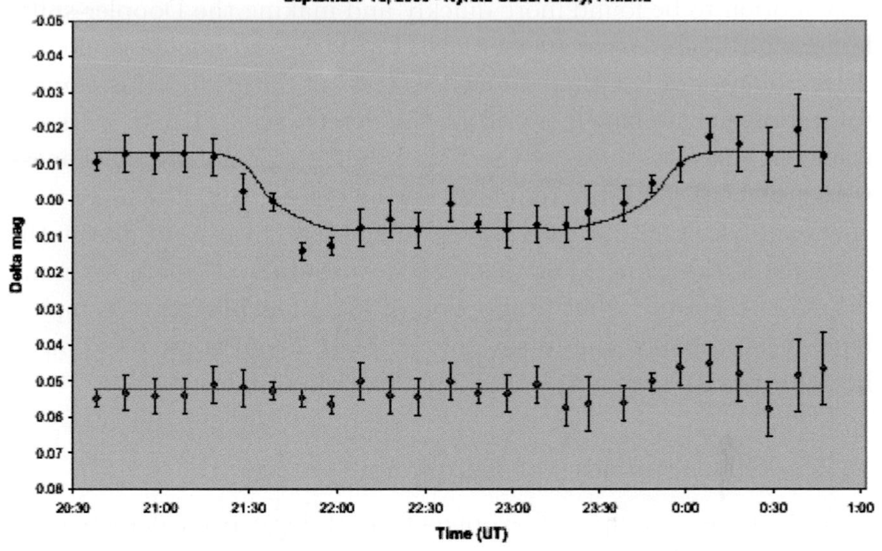

Figure 15.6. The light curve from the transit of an exoplanet over the star HD 209458. The lower line is from a check star. This observation was made by a group of amateur astronomers in Finland at the Nyrölä observatory.

stars are known from their temperatures it is possible in this way to estimate the radius of the planet.

In the description of the determination of planetary orbits and of minimum planetary masses the assumption has been made that the orbits are circular rather than being general ellipses. If the orbit is an ellipse then the curve fitted to the observed stellar speeds is a somewhat distorted version of that previously shown in Figure 15.4 and the degree of distortion enables the eccentricity of the orbit to be found.

The number of known planets around other stars, *exoplanets*, increases steadily with time and is likely to do so for the foreseeable future. Not all stars have planets or, more precisely, detectable planets. The easiest planets to detect are those with large masses in close orbits. A large planetary mass makes the motion of the star greater and hence easier to detect. Close orbits give shorter orbital periods and higher orbital speeds thus enabling a complete cycle of the planetary motion to be found more quickly and making the Doppler-shifts larger and hence easier to measure. Early measurements could not detect planets of less than about the mass of Saturn but instrumental improvements have now led to the detection of planets with less than ten times the mass of the Earth. For a star of, say, solar mass with a planetary orbit of radius 30 au the period would be more than 160 years, so it would take several tens of years just to detect the presence of the planet. For this reason we cannot estimate with any degree of certainty what proportion of stars has planetary companions. It is probably safe to say that *at least* 7% of stars do so but a better estimate may not be available for some time.

15.4. Characteristics of Exoplanets

A fairly typical sample of exoplanets that have been detected are given in Table 15.1.

There are several points of interest in this table. The first is that there are a number of orbits that are very close to stars. The nearest planet to the Sun is Mercury at a distance of 0.4 au. The first four entries in the table are between about one-sixth to one-tenth of the

Table 15.1. Characteristics of a sample of exoplanets.

Star	Minimum mass of planet (Jupiter units)	Period (days)	Semi-major axis (au)	Eccentricity
HD 187123	0.52	3.097	0.042	0.03
τ-Bootis	3.87	3.313	0.0462	0.018
51 Peg	0.47	4.229	0.05	0.0
υ-Andromedae	0.71	4.62	0.059	0.034
	2.11	241.2	0.83	0.18
	4.61	1266	2.50	0.41
HD 168443	5.04	57.9	0.277	0.54
16 CygB	1.5	804	1.70	0.67
47 Uma	2.41	3.0 years	2.10	0.096
14 Her	3.3	1619	2.5	0.354

Mercury distance. A second point is that υ-Andromedae has a *family* of planets, at least the three that have been detected. When there is more than one planet, say three, the star velocity values have to be fitted to a complicated curve which is the sum of three simple curves with different periods and different amplitudes[1]. A final point of interest is that some planetary orbits have high eccentricity, much higher than those in the Solar System. The orbits of highest eccentricity in the Solar System are those of Pluto and Mercury, which are 0.249 and 0.206 respectively. The highest eccentricity in the table is 0.67 and there are some exoplanets with even higher values.

All these observations of exoplanets are relevant to theories of the origin of the Solar System. It would be extraordinary if the other planets were formed in some way different to the way that solar-system planets were formed. If that is accepted, then any viable theory for the origin of the Solar System must, perforce, also be able to explain exoplanets and their characteristics.

[1] The amplitude of the curve is one-half of the difference between the maximum and the minimum.

Chapter 16

Disks Around Older Stars

And the stars are old...

Bayard Taylor (1825–1878)

16.1. The Sun has a Disk

People in the well-developed regions of the world, particularly in or near urban areas, live in an environment of light pollution. The light generated by modern society is scattered by the atmosphere and creates a low level background against which heavenly bodies are seen. This has little effect on viewing bright objects, such as planets or most stars, but fainter objects are blotted out by this background light. In a remote area of the world, say in the midst of a desert, the view of the sky is remarkably different. There are many more stars to be seen with the naked eye than urban man could possibly see and faint features stand out crisp and clear. One such feature is a band of light in the sky called the *zodiacal light*.

The plane in which the planets move is heavily populated with dust varying in size from tiny microscopic particles through to the sizes of grains of sand and some even larger than that. Sunlight falling on these particles is scattered, most strongly at small angles, and it is this scattered light that constitutes the zodiacal light.

Of course, it is necessary for the Sun to be below the horizon for the zodiacal light to be seen. Light is also scattered reasonably well in a backward direction, i.e. towards the source, and so it is also

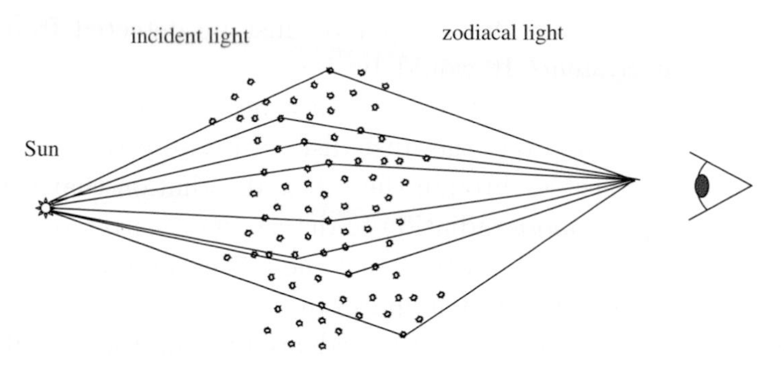

Figure 16.1. Sunlight scattered by dust near the orbital planes of the planets gives the zodiacal light.

possible to detect backscattering of light from a direction opposite to that of the Sun. This is called the *gegenschein* or *counterglow*. From these two phenomena we deduce that there is dust situated on and near the plane of the planetary orbits both inside and outside the Earth's orbit.

One well-known effect of the Sun's radiation on small orbiting particles, known as the *Poynting–Robertson effect*, is that it draws them into the Sun. A grain of sand orbiting near the Earth would be sucked into the Sun in far less than one million years. The time is longer for larger particles and for those further out but the conclusion is inescapable that the presence of these particles is only feasible if there is a source for producing them that compensates for their continual loss. There are, in fact, two main sources. One is the erosion of comets that takes place especially during their passages close to the Sun. The space traveller Regayov (Chapter 5) noted that the tails of comets contained both gas and dust. This dust follows the comet in its orbit and after a considerable time occupies the whole orbit of the comet in a rather non-uniform way. When the Earth goes through one of these dust streams, which it does from time-to-time during the year, we then see a display of *meteors* or *shooting stars*. The other source, also suggested by Regayov, is the occasional collisions of asteroids that not only produces larger bodies, known as meteorites when they land on Earth, but smaller particles as well.

How would this very fine carpet of dust be detected from outside the Solar System? It would be impossible to see, in a visual sense, since so little sunlight would be scattered in the direction of a potential viewer. However, the dust disk — because that is what it is — might be detected through the infrared radiation that it emits. Although this gives more radiation to detect than scattered sunlight, it would still be very difficult to detect the infrared radiation because the Sun's disk contains so little material.

One final point should be stressed about how the visibility of material is linked to the form in which it exists. A planet such as the Earth could not be detected through its infrared radiation by observations taken from, say, a distance of ten light years. However, if the substance of the Earth was converted into dust and spread as a carpet over the whole plane of the Solar System then it could easily be observed. It is all a question of how much surface area is available to emit the radiation. The surface area of the Earth is about 40 million square kilometers. If its mass were converted into sand-grain size then the total surface area of all the particles would be ten thousand million times greater!

16.2. Disks Around Other Older Stars

There are a number of stars, younger than the Sun but by no means *new*, where dust disks can be detected through the infrared radiation they emit. The density of these disks is one thousand or more times that of the Sun, which explains why they can be detected. The first question, that is raised merely by the fact that a disk exists, is to ask what it is that maintains it since the validity of the Poynting–Robertson effect is not restricted to the Solar System.

Figure 16.2 shows an image of the disk around the star Beta Pictoris, where the infrared information has been transformed electronically to make a visual image. The centre of the image, corresponding to the star itself, has been blocked out otherwise the radiation from it would swamp everything else. The disk is seen obliquely, hence its oval shape, and it has a radius of several hundred astronomical units (au). The age of Beta Pictoris is a few hundred million years, enough

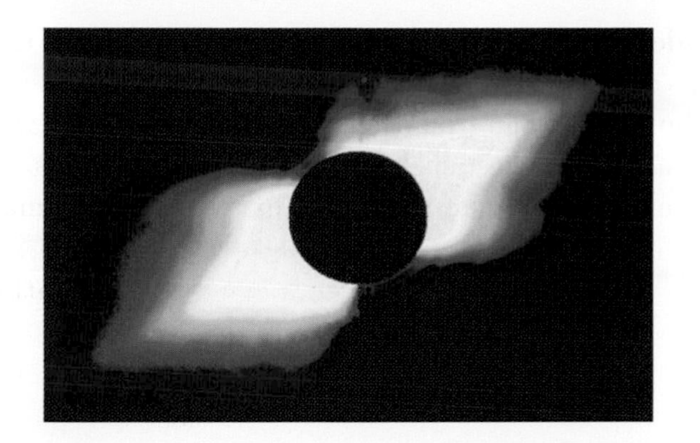

Figure 16.2. An image of the dust disk around Beta-Pictoris.

to cause all but the very largest debris to spiral into the star, so there must be some rich source of dust within it.

There are other younger, but not new, stars for which substantial dust disks have been observed — notably Vega, Fomalhaut and Epsilon Eridani. All the dusty disks have features that suggest the presence of larger bodies. The middle region of the disk of Fomalhaut is cleared out, suggesting the presence of planets absorbing the dust. The brightest part of the disk is at a distance of about 40 au and it has been suggested that what is being seen is emission from dust released by a ring of comets surrounding the star.

Vega gives an even more intriguing observation. Once again, the emission is not close to the star but there is a brightness concentration in a peak at a distance of about 80 au. The origin of this peak is uncertain but one explanation offered is that it comes from a dust cloud surrounding a major planet. Such planets could not be detected by Doppler-shift methods because of the large time required for observation. The emission from Beta Pictoris is mainly concentrated around the star but, once again there is a bright blob that may indicate a planet at several hundred au from the star. Epsilon Eridani shows all the features mentioned — a cleared centre, a ring and a bright spot in the ring that may indicate the presence of a planet.

The evidence is seductive but not conclusive. There is the possibility, perhaps even the strong possibility, that planets exist at distances of tens to hundreds of au from their parent stars. If they were accompanied by smaller bodies, such as comets and asteroids, then these could be the sources of dust that must exist in order to maintain the disks.

Chapter 17

What a Theory Should Explain Now

It has long been an axiom of mine that the little things are infinitely the most important.

Conan Doyle (1859–1930), *Copper Beeches*

17.1. The Beginning of the 21st Century

Although the hope has often been expressed that exploration of the Solar System by spacecraft would lead to new understanding of how the system began, this has not happened. Naturally our knowledge of the Solar System is immeasurably greater than it was in 1960 but that knowledge has not helped greatly in the task of finding a theory of origin. For example, it is of great interest to know that the Jupiter's satellite Io has active volcanoes but it does not indicate how that body was formed. Probably the most helpful information we have had from space research is of a chemical nature. The chemical similarities and differences between lunar rocks and those on Earth might perhaps be saying something about the origin of the two bodies, but might equally be indicative of events that happened after the bodies were formed. This is a constant problem in trying to interpret the results of space research. The Solar System has been in existence for four thousand five hundred million years and it has been an eventful period. We can see on Earth, and on other bodies, the results of bombardment with large projectiles but other cataclysmic events may have taken place, leaving clues that are difficult to decipher.

The astronomical observations made in the last ten years of the 20th century have been far more helpful. The observations of exoplanets enable us to rule out of contention any theory that would make the Solar System a unique example of its kind. A type of event of low probability — for example, as postulated by Jeans and Jeffreys — can be discounted on purely statistical grounds without considering all the detailed criticism of the mechanisms they proposed. While the evidence from dusty disks around older stars is not quite as strong, any theory that could not produce planets at distances of hundreds of au might have a cloud of uncertainty hanging over it.

The Solar System has many complex features and it is expecting a great deal of any theory that it should be able to explain them all. We should distinguish between those features that are mandatory for any theory to explain and those that can be put aside as possible evolutionary features. There may be differences of opinion in the categorization of features. Nevertheless, there will now be given a selection of features that should be addressed by any plausible theory of the origin of the Solar System.

17.2. The Sun and its Properties

Although the search for possible theories is not some kind of competition it still seems a little unfair that monistic and dualistic theories are not presented with the same challenges. Inevitably, a monistic theory, that is, producing the Sun and the planets from the same pool of material, will be required to give a slowly spinning Sun. Dualistic theories assume a pre-existing Sun and duck the issue of slow spin. There is the point that slowly spinning solar-type stars are a matter of observation and a dualistic theorist will clearly use that argument. Nevertheless, to go some way in establishing a level playing field, any dualistic theory that is accompanied by some ideas about the Sun's origin that could explain its slow solar spin should be given brownie points. Equally, a monistic theory *must* convincingly explain the slow spin of the Sun.

The 7° tilt of the solar spin axis could conceivably be an evolutionary feature but extra brownie points should be given for an explanation of this property.

17.3. Planet Formation

This might seem a rather trite and obvious feature for a successful theory but achieving the formation of planets has proved to be quite a challenging theoretical exercise. Because of the lesson learned from Jeans theory, the planets must be made from cold material, where the term cold is to be judged by stellar standards rather than by the standards of the kitchen.

The two kinds of planets, terrestrial and major, occupy different regions of the Solar System. It is possible that, for example, planets are formed as gas giants and later evolve into terrestrial planets, in which case division into the two types is an evolutionary feature. Some theories may naturally give both types of planet but for those giving only gas giants some explanation of terrestrial planets would obviously be desirable if not essential.

The model for planetary formation must be capable of explaining exoplanets as well as those of the Solar System so the mechanism must be a robust one that does not depend on fine tuning the parameters of the model — e.g. temperature, density, etc.

The orbits of the planets must be direct and nearly, but not quite, coplanar.

17.4. Satellite Formation

There is no evidence that exoplanets possess satellites so it may not be necessary for satellite formation to be an essential requirement of a general theory. However, satellites are so common in the Solar System that some generally applicable mechanism should be required for producing them in the Solar System, if nowhere else.

The space traveller Regayov identified some satellites as probable captured bodies. For the most part, these are small satellites, and if captured, can be regarded as of evolutionary origin. The same may also be true for the Moon because of its size relative to that of the

Earth. However, an evolutionary origin for the larger satellites of the major planets would *not* be expected, so some model of their formation is desirable.

17.5. Asteroids and Comets

Some theories have asteroids and comets as precursors of planet formation. For such theories it is therefore required that the origin of these basic bodies should be explained. Other theories make planets directly from diffuse material without the intermediate stage of producing smaller solid bodies. Again, in the interest of having parity of treatment for two approaches, the latter theories should be considered more favourably if in addition to giving a plausible model for planet formation, some process for formation of the smaller bodies is also given.

17.6. Concluding Remarks

There has been such a variety of theories proposed over the years, with different assumptions built into them, that the facts their authors try to explain are also very variable. A prominent American astronomical historian, Stephen Brush, concludes, perhaps cynically, that every individual's list of facts to be explained consists of just that set of features that his or her theory *can* explain. This, after all, is the basis of the quotation from Karl Marx that heads Chapter 6. There may be an element of truth in Brush's comment but everyone would agree that the more that can be explained the better, and that there are some features of the Solar System that are so basic, e.g. planet formation, that they must be on everyone's list.

THE RETURN OF THE NEBULA

Chapter 18

The New Solar Nebula Theory: The Angular Momentum Problem

I shall return.

General Douglas MacArthur (1880–1964)

18.1. A Message from Meteorites

Studies of meteorites in the 1960s indicated that many of their characteristics could be interpreted in terms of their origin as condensations from a hot vapour. We are all familiar with pictures of molten rocks pouring out of a volcano. Now, we have to think of even higher temperatures, where the rocky material does not just melt but turns into a vapour — just as water turns into steam when it boils. If the water was boiled in a room where the prevailing temperature was well below freezing point, then when the steam cooled, it would turn directly into small particles of ice. So it is for the rock vapour. When it cooled it would change directly into small rocky particles that could later be compressed to form larger solid bodies.

There is, however, an important difference between the water vapour and the rock vapour. The description of water as H_2O is familiar — one atom of oxygen linked to two atoms of hydrogen. Water vapour consists of large numbers of H_2O units all moving around freely in space and when they condense to form either water or ice they do so with no change of identity. By contrast, when a rock is vaporized it breaks up into units that are smaller than the chemical units that constitute the minerals in the rock. For example, the mineral fosterite, a form of olivine that is an important major component of the Earth's mantle, has the chemical formula Mg_2SiO_4 — a combination of two atoms of magnesium, one of silicon and four of oxygen. However, there are no identifiable discrete units in the mineral, separated from other units, corresponding to this formula. Rather, the atoms are arranged in a three-dimensional framework and the chemical formula describes the contents of a basic unit of the framework that, when packed together, creates the whole mineral. Vaporizing the material breaks up the framework to produce small very stable entities that then become the components of the vapour. Thus fosterite will break up as

$$Mg_2SiO_4 \rightarrow 2MgO + SiO_2$$

where chemically the right-hand side is two molecules of magnesium oxide and one molecule of silicon dioxide (silica). When the rock vapour is cooled, the individual units can reconnect to form minerals that are different from the original minerals that produced the vapour in the first place. Just what new minerals are produced depends on the rate at which the rock vapour is cooled and also on how diffuse or concentrated the vapour is. In general, as the rock vapour cools so the first minerals to condense out are those that vaporize at the highest temperatures followed by those that vaporize at progressively lower temperatures. Many of the characteristics of meteorite minerals are explicable in terms of 'condensation sequences' from a hot vapour. Another clue is that small crystals within cavities in meteorites showed all the characteristics of having been deposited directly from a vapour.

The message was clear — at some time in the history of the Solar System, the material from which at least some meteorites formed had been in the form of a vapour. This stimulated the thought that this vapour could have been a nebula of the type first envisaged by Laplace. It was realized that Laplace's model had floundered on the angular momentum problem, mainly manifest in the slow spin of the Sun. But, it was argued, more was known in the twentieth century than in the eighteenth so that theorists were confident that somehow or other the problem could be solved. The nebula had returned and the development of the new Solar Nebula Theory (SNT) had begun.

18.2. Mechanical Slowing Down of the Sun's Spin

The first burning problem to be tackled was just how to transfer angular momentum from the core of the collapsing nebula, which would eventually become the Sun, to the disk of material left behind within which the planets would form. The first idea in this direction came in 1974 from the Cambridge astronomers, Donald Lynden-Bell and John Pringle. There are various reasons why the nebula would not collapse in a well-ordered quiet and smooth way. The very process of collapse could introduce some haphazard motion into the nebula. Again, if the core became hot then the disk could be stirred up by being heated just as water in a heated saucepan becomes agitated. Another possibility is that material from outside falls onto the disk, creating disturbances in the way that a pond is disturbed by a stone thrown into it. In summary, the collapsing cloud is likely to become turbulent, a condition that is described more fully in Chapter 26. In a material with turbulence, mechanical energy gets turned into heat and this heat can then be radiated out of the system and lost. The net effect is that in a turbulent medium the energy of motion steadily reduces.

In Chapter 7, where the idea of angular momentum was first introduced, there was given the important principle that in an isolated system, angular momentum remains constant. So it is for the collapsing nebula — the angular momentum must remain constant, but because of turbulence, the energy of motion must reduce. How does the material rearrange itself to satisfy these two conditions? This was the problem solved by Lynden-Bell and Pringle. They showed that material on the inside of the disk must move further inwards while material on the outside must move further outward (Figure 18.1). This means that inner material loses angular momentum while outer

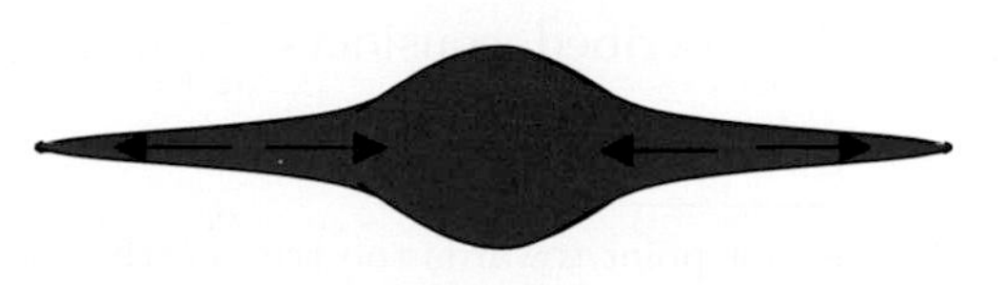

Figure 18.1. Edge view of the nebula showing the motion of material due to turbulence.

material gains it, which is tantamount to an outward transfer of angular momentum from the inner collapsing core to the disk — just what is required.

This process is helpful but it does not solve the problem completely. The way that the inner material moves is by gradually spiralling inwards. At all times it is orbiting at a speed that corresponds to circular motion around the inner mass. When it joins the central mass it would be in free orbit at the equator just as though it was an orbiting planet. A body formed in this way would nearly, but not quite, be at the point of spinning so fast that it would fly apart. That is nothing like the state of the present Sun so there must be some other mechanism available to transfer more angular momentum from the core to the disk.

18.3. Magnetism Gives a Helping Hand

The majority of large astronomical bodies, from the size of planets upwards, have associated magnetic fields. The magnetic field of the Earth shows itself by its action on a compass needle that always points northwards[1], which has been a navigational aid since the time of ancient China.

Figure 18.2 shows a simplified picture of the Earth's magnetic field with the lines, called *flux lines*, giving the direction of a compass needle at various points in the vicinity of the Earth. Of course the flux lines are not physical realities — they are a conceptual construct that enables us to model the behaviour and action of a magnetic field. The north ends of the little bar magnets in the figure are white.

The planet Jupiter has a larger magnetic field than that of the Earth and the Sun's magnetic field is even larger than that of Jupiter. The magnetic fields associated with newly-formed stars can be even greater than that of the Sun and the model for the outward transfer of angular momentum, now to be described, considers a young rapidly-spinning

---

[1] A compass needle does not point towards the true north pole but towards the *magnetic north pole* that moves around but is now situated in the far north of Canada.

Figure 18.2. Magnetic flux lines for the Earth.

star with a magnetic field, accompanied by a disk. A mechanism for transferring angular momentum by a magnetic field was suggested by the British astrophysicist Fred Hoyle (1915–2001) as early as 1960. His idea depended on having the central collapsing core and the edge of the disk at a temperature high enough for the material to be ionized. The ionization process means that atoms are broken up into negatively charged electrons and the rest of the atoms (called ions) with a positive charge. An intimate mixture of electrons and positively charged ions is called a *plasma* and, because they consist of charged particles, plasmas are good conductors of electricity.

Hoyle

If a magnetic field exists within a highly-conducting plasma then it will be *frozen in*, meaning that if the plasma moves then the flux lines will move with it. Hoyle envisaged the central collapsing core of a nebula separated by a gap from the inner edge of the surrounding disk, also consisting of ionized material, with magnetic flux lines linking them (Figure 18.3).

The flux lines behave like rubber bands and if they are stretched they try to shorten themselves. Those shown in the figure are frozen into both the core and the disk meaning that they are firmly attached at each end. The core spins more rapidly than the disk so it stretches the flux lines. In trying to shorten themselves, the flux lines pull inwards at each end and from the figure it is clear that this will have the effect of slowing down the spin of the core and speeding up the spin of the disk — effectively transferring angular momentum outwards.

The problem with the Hoyle model is that the magnetic field that he assumed was too small, so that the field lines became greatly stretched before they could exert a significant force on the core and disk. It was as though the rubber bands were rather thin and easily stretched. Theory shows that, before the flux lines stretched to the

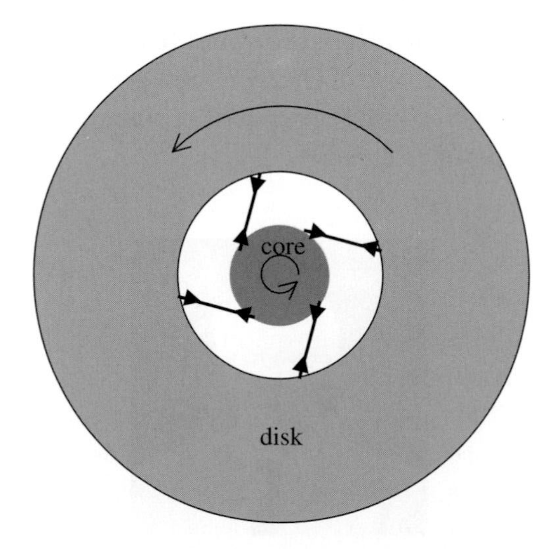

Figure 18.3. Flux lines linking a collapsing core to a disk.

extent required, they would break and the core and disk would then reconnect with new shorter flux lines. The flux lines would be continuously stretching and breaking and there would never be sufficient force acting on the core and disk to transfer angular momentum.

18.4. A Modification of the Hoyle Mechanism

There have been several newer ideas about transferring angular momentum outwards by the use of a magnetic field that differ in the initial assumptions they make. Here we describe one that has been worked out in some detail by Armitage and Clark in 1996.

The core is taken to be in the form of a very active pre-main-sequence star with a very high magnetic field, perhaps one thousand times that of the Sun. The disk is in orbit around the central mass with material in the inner part of the disk orbiting at a rate faster than that of the core. The disk material further out is orbiting at a lower rate, just as outer planets orbit the Sun more slowly than the inner ones. In Hoyle's original model, the magnetic linkage was just to the inner region of the disk, but now magnetic linkage is assumed to occur between the core and all parts of the disk, as illustrated in Figure 18.4 in edge view. The magnetic flux lines linking the core to the more slowly rotating outer material transfer angular momentum outwards just as Hoyle had described. Consequently that material moves outwards. Because the magnetic field is so strong, the flux lines behave like very strong rubber bands that have to be stretched very little to give the required forces and therefore do not break and reconnect. The more rapidly orbiting material in the inner part of the disk is also connected to the core by flux lines but these give a different form of behaviour. The inner part of the disk is now assumed to become rather turbulent and material from this part of the disk flows along the flux lines to join the central core.

The transfer of angular momentum through the action of a magnetic field requires several conditions to be satisfied. The core must generate a field of sufficient strength and although strong surface fields have been detected on some young stars, it is not completely certain that the field would be of a form that would enable a strong

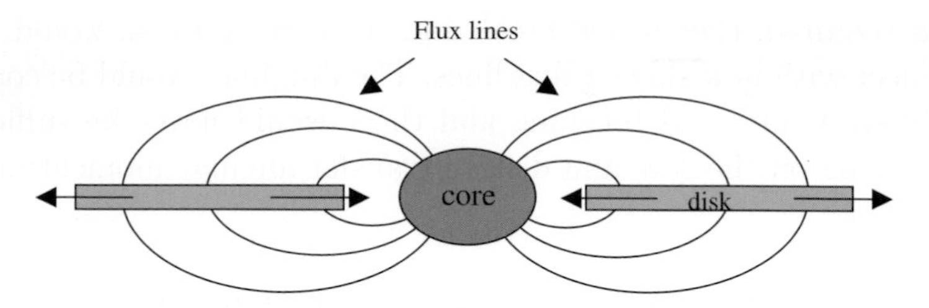

Figure 18.4. Inner disk material moves inwards to join the core while outer disk material moves outwards.

linkage to occur. Another requirement is that the disk material should be hot enough to be conducting at some considerable distance from the star. That should not be a problem at the inner edge of the disk, which is close to the star and fully exposed to its radiation. Finally, the disk would have to be reasonably quiet and not too turbulent as otherwise, it may not be possible to set up the required pattern of flux linkage. Another problem with the particular model described here is that the initial state taken for the core was far from being at the point of rotational disruption, which we have already seen is what happens through the action of the Lynden-Bell and Pringle mechanisms. The important process of attaining this initial state has still to be described.

18.5. Slowing the Sun's Spin

Even if the transfer of angular momentum outwards can be achieved, then it is not very likely that the transfer process would leave the Sun spinning at its present very slow rate — with a period of about thirty days. We now consider another mechanism depending on a solar magnetic field that would slow down the spin of the Sun without reference to any external body such as a disk. This relies on the interaction between the solar magnetic field and the *solar wind*, a stream of energetic particles — mostly electrons and protons — leaving the Sun at a speed of about 1000 km s^{-1}. The rate of loss of mass of the Sun due to the solar wind is two million tonnes per second; to put it in perspective this rate of loss would cause the Sun

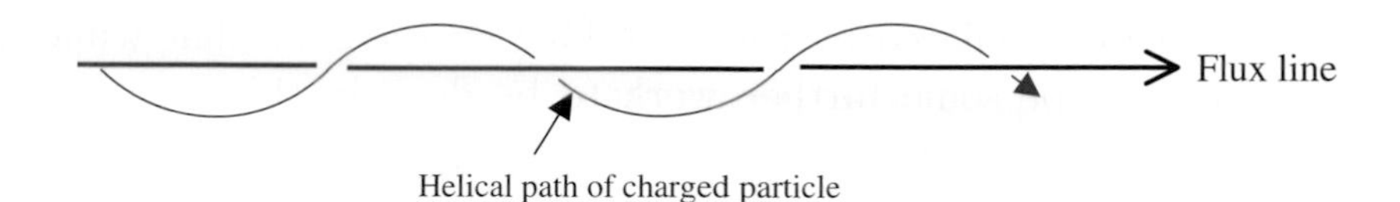

Figure 18.5. The helical path of a charged particle coupled to a flux line.

to lose one ten-thousandth of its mass over its 10,000 million year lifetime.

When streams of charged particles move in the presence of a strong magnetic field they can become coupled together so that the charged particles move along a helical path with a flux line as an axis of the helix (Figure 18.5).

When the Sun, or indeed any star, spins the magnetic field also spins like a rigid structure, and this would apply to all the imaginary flux lines that define the field. This means that the solar wind particles would move outwards along a flux line at a *constant angular speed.* How far the particle moves along the flux line before it breaks free depends on the energy of the particle and the strength of the magnetic field. The more energy the particles have, the sooner they break free; the stronger the field, the later they break free. The consequence of this process is clear from equation (7.1). The distance from the star (r) increases but the angular speed (ω) is unchanged, which means that the angular momentum associated with the escaping particles is increasing. However, the total angular momentum of the system has to be conserved so there is a compensating decrease in the angular momentum of the remaining bulk of the Sun — i.e. it spins more slowly.

With the present rate of loss of matter due to the solar wind and the present strength of the Sun's magnetic field, the effect is a weak one and would not have removed much angular momentum over the lifetime of the Sun. However, it is likely that the early Sun was much more active than it is now and, for a period of a million years or so, the rate of mass loss could have been more than one million times the present rate and the field one hundred times greater than at present. That being so, it is possible that the Sun retains as little as 1% of its original angular momentum.

It is generally believed by those working on the SNT that while there may still be some further work to be done in this area the essential problem of angular momentum transfer can be solved by the action of magnetic fields and that the required starting point of a relatively slowly spinning core and a surrounding extensive disk, containing most of the angular momentum of the system, can be achieved. If that is so, then the next problem to be considered is that of forming planets in the disk.

Chapter 19

Making Planets Top-Down

A great while ago the world began.

William Shakespeare (1564–1616), *Twelfth Night*

19.1. A Massive Disk

While the tidal model put forward by Jeans did not stand the test of time, the theory that underpinned the model is completely valid. For example, a gaseous filament would break up into a string of blobs and any blob that was greater than a certain critical mass would collapse to form a condensed body. There was nothing wrong with Jeans' theoretical work — the failure of his model was due to the scenario in which he applied it. The kind of behaviour that Jeans found for a filament would also occur in a spread-out thick disk of material. It too would be gravitationally unstable and break up into a number of blobs, in this case distributed over the disk rather than strung out in a line (Figure 19.1). If these blobs had greater than the critical mass they could form condensed bodies.

The first idea about making planets with the SNT was through such a process. The disk was taken as having sufficient density and mass per unit area to produce blobs that would be above the critical mass and condense to form major planets. Equation (9.2) shows that the critical mass depended on both the density and temperature of the material. The higher the density the *lower* is the critical mass, the higher the temperature the *higher* is the critical mass. In order to produce critical masses equal to that of, say, Jupiter with material at a temperature that could vaporize silicates required absurdly high

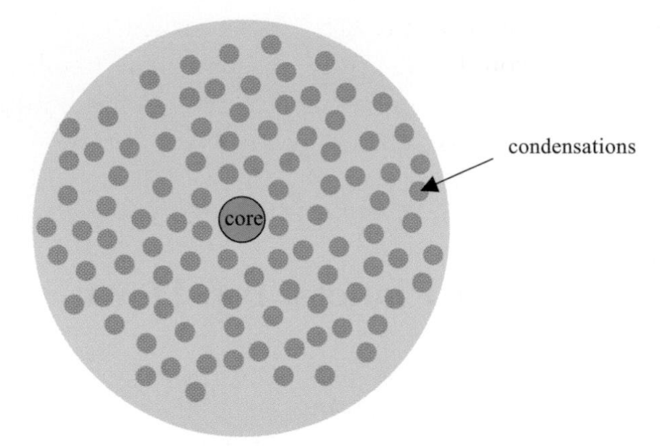

Figure 19.1. The spontaneous break-up of a disk into protoplanets.

densities — so high that the mass of the disk would be very much greater than the mass of the central core that was to form the star. As a result of this the idea of a very hot nebula had to be abandoned and consequently the whole *raison d'être* for the resurgence of nebula ideas was lost. However, by this time the SNT had become firmly established as the most plausible model for forming the Solar System and work on it continued unabated. Other ways would have to be found for producing the vapours that meteorites required — but that was not an immediate concern of the solar-nebula theorists who were more engrossed in the problem of producing planets.

This process of producing major planets by direct condensation of the disk became known as 'top-down'. The idea behind this name was that there is a hierarchy of sizes of objects in the Solar System — dust, small asteroids, large asteroids, satellites, terrestrial planets and major planets, with the major planets at the top of the list. Top-down formation implies that the largest bodies, the major planets, are first produced and the formation of other smaller bodies is through their subsequent break-up in some way.

19.2. Some Problems of Top-Down Processes

Even with a low temperature disk, say in the range 10–100 K, the density required to produce major planets was still so high that the disk had to have a mass that was similar to that of the core. When the disk fragmented, it would have produced not a small number of major planets but a very large number of such bodies. If the conditions were right in one part of the disk to produce a major planet then they would also be right in most other parts of the disk. This raised the question of disposal since a great deal of energy is required to expel the surplus material, probably a large fraction of the mass of the Sun, from the Solar System. A possible source for such an amount of energy could not be imagined.

Later, when disks were detected around young stars, the observations did not indicate the existence of such massive disks; the maximum mass of a disk is thought to be about 0.1 $M_\odot$[1]. Support for the top-down approach declined and other ways of producing planets from a disk were considered.

[1]The symbol $\odot$ indicates the Sun so that $M_\odot$ represents the mass of the Sun.

Chapter 20

A Bottom-Up Alternative

Tall oaks from little acorns grow.

David Everett (1769–1813)

20.1. A Summary of the Bottom-Up Approach

The antithesis of the top-down process is to start the process of forming planets, and other bodies, with the dust contained within the disk. This would have constituted about one percent of the mass of the disk and it would have been in the form of very tiny grains. The proposed stages of planet formation in this way are as follows:

(1) The dust gradually settles down into the mean plane of the disk to form a thin carpet.
(2) Just as for the gas disk described in Chapter 19, the dust disk breaks up into a large number of small compact solid bodies. Following the notation of Chamberlin and Moulton (Chapter 8) these bodies are referred to as *planetesimals*.
(3) The planetesimals come together to form larger solid bodies. In the inner part of the Solar System these are the terrestrial planets. In the outer part of the system they are the cores of the major planets.
(4) The major planets acquire gaseous envelopes that, at least for Jupiter and Saturn, account for most of their masses.

Each of these stages will now be described.

20.2. Forming a Dusty Carpet

The endpoint of this stage of the bottom-up process is to produce a solid layer in the mean plane of the disk consisting of the dust that has settled through the effects of gravity, as shown in Figure 20.1. The dust particles in the nebula disk would have been very tiny, with dimensions typically about one micron. They are constantly being buffeted in all directions by the molecules of the surrounding gas which exert on them much stronger forces than those of gravity although the influence of gravity does gradually bring them towards the mean plane. In producing planets from a nebula we have to keep in mind the observed lifetimes of disks around young stars — one to a few million years. Every process in the bottom-up mechanism must take less time than a disk lifetime as must the total time for all the processes together. For an individual dust particle, the settling time would be of order ten million years, so this presents a problem.

A solution to the problem was suggested by the American planetary scientist, Stuart Weidenschilling and his co-workers in 1989. It was suggested that dust particles in space would stick together for a number of possible reasons. To explain the first reason it is first necessary to point out that although the disks around stars have considerable mass, they are by normal terrestrial standards, extremely diffuse. The dense part of a nebula disk would on Earth be designated as an ultra-high vacuum! Consequently, the surfaces of the dust particles are not often impacted by gas molecules and they remain clean and uncontaminated. Experiments in the laboratory show that when two uncontaminated surfaces of the same material are pressed together the atoms on the surfaces can chemically bond together. Such *cold welding* could occur between clean surfaces of dust particles if they came together.

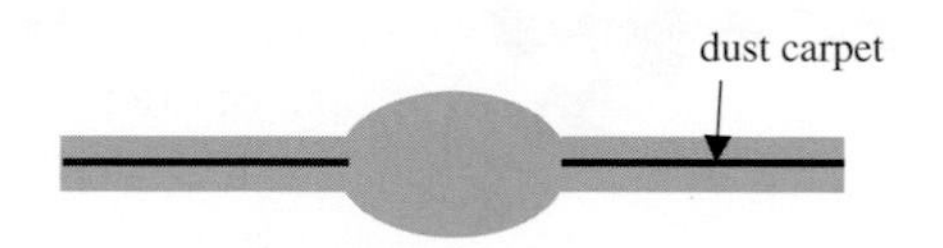

Figure 20.1. The gravitational settling of dust to form a dust carpet.

The second reason why dust particles might join together is through electrostatic forces. This is due to particles becoming electrically charged by friction. Such a process can happen in everyday life with pieces of cling film or tissue paper that can be very difficult to detach from one's hands, or each other, due to the electrical charges they acquire.

A final reason that has been suggested is actually the very opposite of the first one and considers that the dust grains might become covered with some sticky organic materials that would act like glue.

Weidenschilling and his colleagues did some computer simulations of how dust particles would stick together. One of the outcomes is shown in Figure 20.2.

The general form of these aggregates is that they are rather loose fluffy structures. The best way for the dust to come together from the point of view of fast settling would be if it formed very tight compact bodies but according to theory developed by Weidenschilling, even the looser structures as shown would be able to settle in the mean plane on a timescale between one hundred thousand and one million years. This is an acceptable figure in the light of the observed lifetimes of disks.

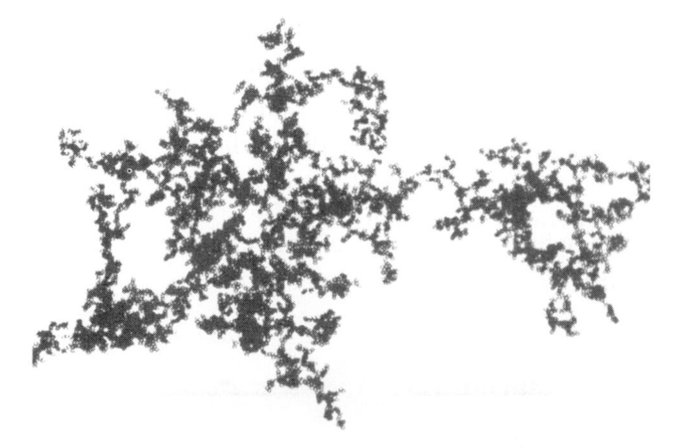

Figure 20.2. Computer simulation of the aggregation of dust particles.

Nevertheless, there is some doubt about this result since Weidenschilling's calculations showed that the settling time is very sensitively dependent on the exact form of the dust aggregates. In 2000, a space shuttle experiment called CODAG (COsmic Dust AGgregation experiment) showed that dust in space did stick together quite readily but not precisely in the form predicted by Weidenschilling's computer simulations. The aggregates were far more strung out in one direction forming a fluffy string rather than a fluffy blob. Because of this, and the sensitivity of the settling time to the form of the aggregates there is some doubt that settling into the mean plane can take place within a disk lifetime. On the basis of the CODAG results, it was suggested that planetary scientists might have to review the way that planets formed but later those working on the CODAG project gave a less pessimistic picture, suggesting that as grains grew to a larger size they became more compact and therefore could settle more quickly. The whole bottom-up process of planet formation depends on this first stage of dust settling but opinions vary on how plausible it is.

Towards the end of 2005, an observation was made of an exoplanet in a triple star system designated as HD 88753. The planet was in orbit very close to a solar-type star. Two somewhat-less-massive stars, with combined mass of 1.63 times that of the Sun, formed a close binary system in an orbit about the single star at an average distance of 12.3 au with eccentricity 0.5. This observation was accompanied by comment that it threw doubt on the bottom-up process of planet formation since the material accompanying the star system would be constantly stirred up, both thermally and mechanically, thus preventing settling of dust into a disk. A possible solution to this difficulty is that the planet formed in association with the solar-type star before the binary pair attached itself — although this explanation introduces new, if different, problems.

20.3. The Formation of Planetesimals

Once a dust carpet formed it would be a very substantial entity. The total mass of the disk might be one-tenth that of the Sun (Sun's mass $= 2 \times 10^{30}$ kg) and if one percent of that was dust then the mass of dust would be about the same as that of Jupiter, 2×10^{27} kg. This mass, spread uniformly in a disk reaching out to the orbit of Neptune would give an average mass per unit area of about 30 kg per square metre. It was shown by the American cosmogonists, William Ward and Peter Goldreich, in 1973 that the dust disk would break up into a vast number of condensations, just as illustrated for the total disk in Chapter 19. Because the dust disk would not be uniform, and also because of different distances from the Sun, the condensations would be of different sizes and masses in different locations. These condensations are the planetesimals, with typical dimensions in the range between hundreds of metres and tens of kilometres.

This part of the bottom-up process has a sound theoretical basis and it is fairly certain that if a dust carpet formed, then planetesimals would be an outcome.

20.4. Making Terrestrial Planets and Cores for Giant Planets

Another model that required planets to form from very diffuse material was the accretion theory suggested by Otto Schmidt in 1945 and later developed by Lyttleton, as described in Chapter 10. Following this original idea, the theory of the first stage of forming planetesimals from dust was developed by two Russian workers, but because of the poor dissemination of information from Russian-language publications, this work was largely unnoticed outside Russia. The following process — that of building up larger bodies from planetesimals — was first tackled by Victor Safronov (1917–1999), who worked at the Institute of Earth Sciences in Moscow of which Schmidt was the Director. Safronov's work was published in Russian in 1969 but an English translation became available in 1972 and this idea soon became central in the development of the SNT.

Safronov

When planetesimals first formed by fragmentation of the dust carpet, they would have moved in close-to-circular orbits. Occasionally pairs of planetesimals would approach each other and through their mutual gravitational attractions they would be deflected out of their circular paths. Consequently, as time progressed, so the pattern of smooth circular motion would break down and the motions of the planetesimals would take on a somewhat chaotic character. As this happened, a new kind of process increasingly came into play. The chaotically moving planetesimals sometimes collided and the effect of this was to damp down the chaos in the motion. Eventually, a state of balance was reached between gravitational interactions increasing chaotic motion, and collisions decreasing chaotic motion, after which the overall degree of chaos in the motion remained more-or-less constant. Safronov showed that when this happened the relative speeds of planetesimals was about equal to the *escape speed* from the largest planetesimal.

The concept of *escape speed* plays an important role in many aspects of astronomy. If a ball is thrown into the air, it comes down again but the faster it is thrown upwards, the higher it goes before it returns. In the period after the Second World War, rockets that had been developed for military use were used as research tools by firing them into the upper reaches of the atmosphere to collect information beyond the reach of balloons. The faster the rocket was launched the further up it would go but there was a limit beyond which a single stage rocket could not go because of the relationship between the mass of its fuel and the energy it contained. Later, two and three-stage rockets were developed where the first stage provided a high platform for the firing of a smaller second stage, and so on. In this way rockets were launched that could escape the Earth's gravity and so become the carriers of space vehicles. For a spherical body of mass M and radius R the escape speed from its surface is given by

$$v_{esc} = \sqrt{\frac{2GM}{R}} \tag{20.1}$$

The escape speed from the Earth is about 11 km s^{-1}.

If two bodies move together and collide under the influence of their mutual gravitational attraction then, from theory, their relative speed on collision is at least equal to the escape speed of each from the other. If none of the energy of motion is lost then the bodies would simply bounce apart and separate again. However, if energy of motion is lost in the collision process, say in fragmenting one or both of the bodies or converting some energy into heat, then the two bodies can join together to form a single body. This process works better if one of the bodies is significantly bigger than the other when it can be pictured as accretion of the smaller body onto the larger (Figure 20.3).

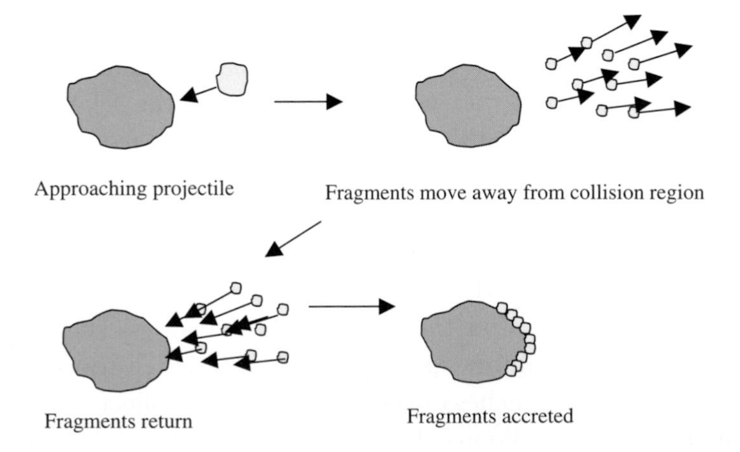

Figure 20.3. The accretion of a small body by a larger body due to a collision.

The more massive a planetesimal is the greater is its attractive force on other planetesimals and therefore the greater the number of other planetesimals it accretes. But, the more it accretes the larger its mass and therefore the greater its attractive force — and so on. The net result of this process is that in any particular region of the disk one planetesimal will grow and become dominant. In the inner part of the Solar System these dominant planetesimals will become terrestrial planets but further out they will become the cores of major planets.

Safronov's analysis was mathematically based and he worked out the rate at which the cores or terrestrial planets grew and the total time for their formation. For example, to produce the Earth would have taken about one million years and the formation of a Jupiter core about two hundred and fifty million years. We now know that the time for producing Jupiter is unacceptable because of the observed lifetimes of disks — but that information was not available to Safronov when he carried out his analysis. The time of formation increased as one moved outwards from the Sun and it was the outer planets that gave the most disturbing results. The time for producing Neptune would have been about ten thousand million years, greater than the age of the Solar System, and Safronov *was* aware of that. This timescale problem was also noted by the SNT community but it was thought that Safronov's scheme could be somehow modified to shorten the formation times.

20.5. Major Planets — The Final Stage

It must now be taken that planetary cores have been produced within the gaseous nebula within a timescale such that a large quantity of nebula gas is still present. Due to the newly formed Sun, for so it may be called at this stage, the temperature of the nebula would have been high in the inner regions and cooler as one moved outwards. In the region of the terrestrial planets the temperature may have been too high, and the solid cores too small, for gas to be retained when it came into contact with the cores. It should be noted that the Sun may have been much more luminous when it was first formed than it is now so that the temperatures in the region of the terrestrial planets may have been much higher than they are now. We should also note that even at the present temperature of the Earth it would be unable to retain a hydrogen atmosphere.

In the outer part of the Solar System, from Jupiter outwards, conditions would have been suitable for the capture and retention of hydrogen and helium and calculation indicates that the acquisition of the gas, that is the majority component of Jupiter, would have taken no more than one hundred thousand years.

There was now in place a complete model for the formation of planets by the bottom-up process starting with dust. Some of the problems of the model were not known at the time it was first formulated and, as problems have become apparent, so work has been done to address them. However, a problem that was evident from the earliest development of the bottom-up process was the long time it would take to form the outermost planets.

Chapter 21

Making Planets Faster

...things can never be done too fast.

Oliver Goldsmith (1730–1774), *She Stoops to Conquer*

21.1. Conditions in the Disk

The process described by Safronov for making planets from planetesimals was certainly a positive step forward. If sufficient time was available then a mechanically based series of steps had been well-described using sound physical principles and with a good mathematical analysis. At first, it seemed that only Uranus and Neptune presented timescale problems but when disk lifetimes had been estimated as a few million years at most, all the major planets were found to have formation times that were too long.

The American cosmogonist, George Wetherill, has considered this problem in some detail. Various conditions within the nebula disk can be proposed, to substantially shorten the Safronov estimated formation times.

21.1.1. *The initial masses of planetesimals*

If the planetesimals in one region were all of similar size then it would take a long time for one of them to become dominant. By chance some would accrete more than others and at first a number of planetesimals would be competing for the role of 'dominant planetesimal'. Eventually just one would emerge and this would be able to collect much of the mass within its region of dominance. However, the process can be sped up if the initial masses are very non-uniform, especially if one planetesimal is much more massive than anything else in its neighbourhood. This could come about if the nebula was somewhat 'clumpy' instead of having a smoothly varying density that just depended on distance from the Sun. A large clump would give rise to a large planetesimal and by this means a locally dominant planetesimal could be present straight away.

21.1.2. *Big bodies move more slowly*

There is a principle in physics that goes under the name 'energy equipartition'. It applies to a system of bodies, interacting with each other in any way, where not all bodies have the same mass. The type of interaction can be gravitational or by collisions — it matters not, the principle will still apply. Basically, it says that the probable energy of motion will be the same for all bodies regardless of mass. Since the energy of motion, known as kinetic energy, increases with both mass and speed this implies that if the mass is higher, the speed must be lower and *vice-versa*. This principle is illustrated in Figure 21.1.

The application of this principle means that the dominant planetesimal will tend to be moving more slowly than the others, and

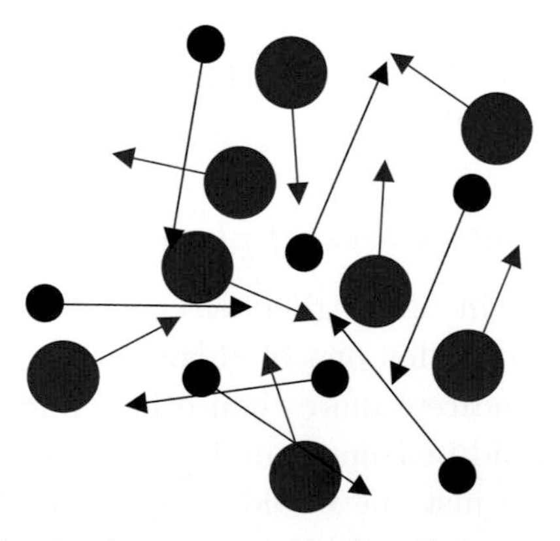

Figure 21.1. In an interacting system of bodies, the more massive bodies move more slowly.

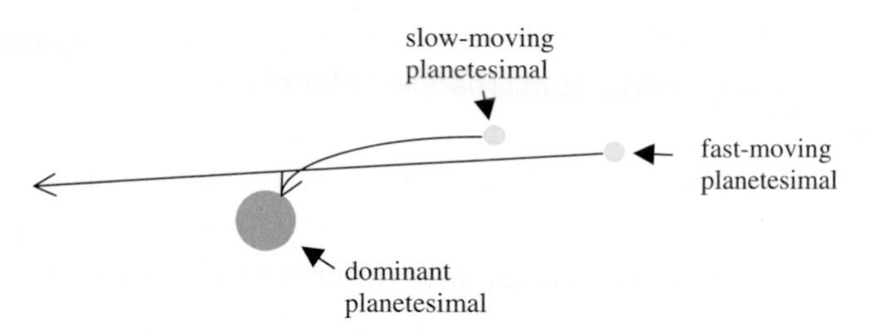

Figure 21.2. The gravitational deflection is greatest for slower moving planetesimals.

of the remaining planetesimals, the larger ones will be the slowest movers. Hence, the most slowly moving planetesimals *relative to* the dominant one will tend to have greater than average masses. Another factor here is that the range of attraction for other bodies of a body with a particular mass increases when the relative speed is less. A slowly moving planetesimal will curl in towards the dominant one whereas a fast moving planetesimal will just pass by without much deflection (Figure 21.2).

For these reasons there will be a tendency for the dominant planetesimal to preferentially accrete the largest of the other planetesimals. This increases the rate of growth over that calculated by Safronov.

21.1.3. *Gas drag*

The makers of motor vehicles and of aircraft are always concerned with the question of air resistance, or drag, the force on the vehicle due to its motion through the air. In the case of aircraft, flying at a great height reduces the density of the surrounding air and hence reduces the drag. This cannot be taken to the ultimate limit of leaving the atmosphere altogether as the presence of some air is required to give the lift that supports the plane. For a car there is no choice of environment; the car must travel through the air at ground level and all that one can do is to design its shape to offer as little resistance as possible.

The planetesimals would have been moving within the gas of the nebula and so will be subjected to drag that will slow down their motion. The drag will actually be more important for smaller planetesimals and any body that reaches planetary size will thereafter experience a negligible drag effect. It is all a question of how much surface is exposed to the gas, which for a given speed is the main factor determining the total force on the body, in relation to its mass. For two bodies of similar shape but differing in linear dimension by a factor of two, the larger body will have four times the area, and hence four times the resisting force, but this will be acting on eight times the mass — hence the slow-down due to drag on the body is less. Since gas drag will be relatively ineffective in slowing down the larger planetesimals, it will be only marginally important for increasing the growth of the dominant planetesimal, but it will have some effect.

21.2. Runaway Growth

Applying all these factors can lead to what Wetherill has called *runaway growth*. It implies that the process by which the dominant planetesimal grows is an accelerating one so that the increase in mass gets faster and faster with time. The growth will slow down and cease altogether when the material is all locally exhausted. An analysis of runaway growth was described by Stewart and Wetherill in 1988.

The estimates of the times of formation of terrestrial planets and cores are drastically changed by runaway growth. A Jupiter core of ten Earth masses can now be formed in about one million years. The times for producing Uranus and Neptune are also much shorter, somewhere in the range from ten to thirty million years. The times for the outer planets are still uncomfortably large but at least they are within a factor of ten of being acceptable.

There are some nagging doubts about the runaway growth model. An assumption in the runaway growth model is that of a high density of planetesimals thus enhancing the rate of accretion by the dominant planetesimal. According to the American cosmogonist Alan Boss, the assumption made by Stewart and Wetherill requires a disk density greater than can reasonably be expected. A further difficulty was raised by Wetherill himself. Once the dominant planetesimal has reached about one Earth mass then it will not only accrete planetesimals but it will also scatter them all over the Solar System. Once they are scattered they are not locally available to be accreted. While it could be argued that there will also be planetesimals scattered *into* the region of the dominant planetesimal, these incomers will be moving at high speed and therefore be more difficult to accrete. Many more will be scattered than will be accreted so the local density of planetesimals that can readily be accreted will be much lower than if no scattering had taken place. The calculated rate of runaway growth assumes that all local planetesimals stay local and are always available to be accreted; the fact that they are not means that runaway growth will not be as fast as theory suggests.

Chapter 22

Wandering Planets

Our souls, whose faculties can comprehend
The wondrous architecture of the world:
And measure every wand'ring planet's course,
Still climbing after knowledge infinite.

Christopher Marlowe (1564–1593)

22.1. The Need for Planets to Wander

Even accepting the validity of runaway growth ideas, the outer planets, Uranus and Neptune, still pose a time-of-formation problem. The difficulty stems from where they are formed. According to the original Safronov theory, the time of formation of a planetary core of a particular mass is proportional to the period of the planet in its orbit and to the inverse of the surface density (mass per unit area of the disk) of planetesimals in the region. The period of Neptune is about fourteen times that of Jupiter and the surface density of planetesimals in the vicinity of Neptune, for any sensible model of the nebula, will be at most about one fifth of that near Jupiter. This means that whatever the formation time is for a Jupiter core, the formation time for a Neptune core of the same mass would be of order seventy times greater. Thus, if a Jupiter core takes one million years to form, then the Neptune core would take about seventy million years. There is a dependence on the mass of the core and the runaway model has extra factors within it so the ratio of times found is closer to thirty than seventy but the general principle applies.

Another problem, knowledge of which is more recent, concerns the exoplanets that have been found very close to stars — as close as 0.04 au. Assuming that these planets are gas giants, which seems likely from their masses, then it is improbable that they could have *formed* where they are now — although they are obviously stable in their present environments. A core in such a situation could not *start* the process of building the planet by retaining local nebula gas that would be at a very high temperature.

The solution to both these problems is to posit that all the planets in question were produced in some favourable position from a time-of-formation point of view — say, between the present locations of Jupiter and Saturn — and later wandered, inwards and outwards, to where they are now. Such movements are referred to as *migration* and the processes for producing migration have been extensively studied.

One mechanism that can be directly discounted is drag due to the viscosity of the gas, which is totally ineffective for planetary masses, although it can have an effect on smaller planetesimals.

22.2. Interactions Between Planets

Although the SNT should produce planets in near-circular orbits, they need not stay that way. Planets act gravitationally on each other and the orbit of a planet produced, say, in the asteroid belt would be unstable over a long period of time due to perturbation by Jupiter. If its orbit became highly eccentric in this way then there could be a close passage of the two planets that would move one of them inwards towards the Sun while the other went out further from the Sun. If Jupiter was considerably more massive than the other planet, then there could be a comparatively small inward change in its orbit while the other planet went out to a great distance (Figure 22.1).

This process alone could not explain the present orbits of Uranus and Neptune. The planet we call Neptune may go out several tens of au from the Sun but its orbit would then be very eccentric and take it back into the vicinity of Jupiter's orbit. Some further process would be necessary for it to end up in a circular orbit of radius 30 au.

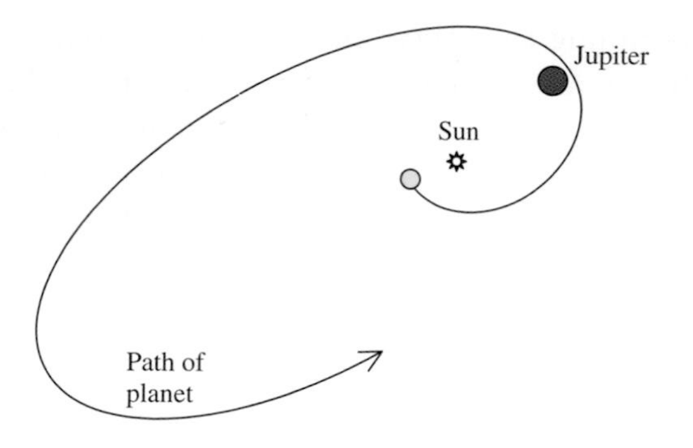

Figure 22.1. A planet thrown outwards by an interaction with Jupiter.

22.3. Effects Due to the Mass of the Nebula Disk

We put off a detailed description of the way that a planet interacts with a disk until Chapter 29 where we consider the effects on more general planetary orbits. At present we just give a description of the evolution of the *circular* orbit of a planet within a disk. It is assumed here that the disk material circulates around the Sun (or a general star) in free orbit under the influence of the central mass so that at any distance from the Sun the material is moving in the same way as a planet would at that point. This means that there is little relative motion between a planet and the medium that is very close to it.

If we consider material well inside, or well outside, the planet's orbit then there will be considerable relative motion. The inner material will be rotating faster and also moving at a higher speed than the planet. Conversely outer material will be moving at a lower speed than the planet, as indicated in Figure 22.2.

Some fairly straightforward analysis (e.g. see reference to Cole and Woolfson, 2002, pp. 459–463) shows that the gravitational interaction between the planet and the slower moving outer material adds angular momentum to the material and takes it away from the planet. The net effect of this is that the material moves further out and the planet moves further in. For the inner material, which moves faster than the planet, the reverse is true. The inner material loses angular momentum and moves inwards while the angular momentum gained by the planet tends to move it outwards. The effect on the material

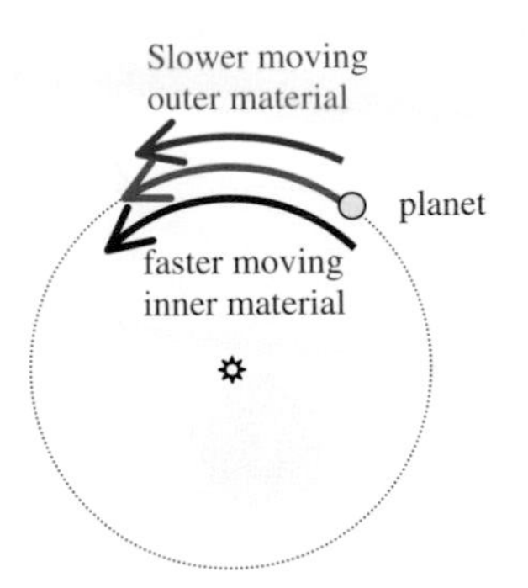

Figure 22.2. The relative motion of the planet with respect to neighbouring disk material.

is straightforward. It moves away from the planet whether it is inside or outside the planet's orbit. What is not quite so straightforward is the effect on the planet that gains angular momentum from inner material and loses it to outer material.

The movements of the medium away from the planet tend to open up a gap within the disk. When this occurs the motion of the planet will be due to *Type II migration*, first described by Ward in 1997. Whether the planet moves inwards or outwards depends on the balance of the effects from the inner and outer material. In all practical situations inward migration takes place far more easily and can take the planet very close to the central star, as is observed for some exoplanets. Outward migration is more difficult to achieve and, when it does take place, the total outward movement is small. Type II migration is not a realistic mechanism to transfer planets formed near Jupiter or Saturn out to the orbit of Neptune.

For a planet with mass less than that of Jupiter, no gap appears so the planet is always in contact with the medium. This gives rise to what is called *Type I migration*. This is caused by the asymmetric effect of wakes forming on either side of the planet that have the appearance of spiral waves. The outer wake tends to push the planet inwards and the inner wake tends to push it outwards but the net effect always leads to inward motion and is extremely effective. Indeed, it is so effective that it seems that almost any planet involved

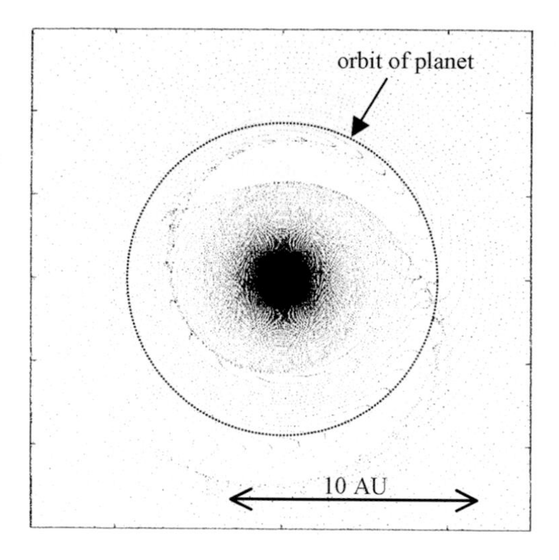

Figure 22.3. The disturbance of the medium by a planet in a circular orbit.

in Type I migration must plunge into the central star. A mechanism that prevents this from happening will be described in Section 22.5.

Figure 22.3 shows the state of the medium under the influence of a planet in a circular orbit found from a simple computer model. Gaps can be seen in the medium although not a clean circular gap as the material is very mobile. Other features worth noting are the spiral structures both inside and outside the planet's orbit.

22.4. The Role of Spiral Waves

The spiral waves seen in Figure 22.3 are carriers of angular momentum. Those moving outwards from the planet's orbit are transporting angular momentum from the region of the planet outwards. As they travel they dissipate, meaning that they give up energy and angular momentum to the medium within which they move. Another way they can dissipate is by impinging on some external planet that then acquires energy and angular momentum. If an outer planet is constantly being fed by these waves generated by a massive interior planet then this is a way of moving the outer planet outwards. Of course, since angular momentum must be conserved the inner planet has to move inwards.

It has been suggested that Uranus and Neptune were moved outwards by such a process, the provider of the spiral waves being Jupiter. Since Jupiter has ten times the mass of Uranus and Neptune combined, a comparatively modest inward motion of Jupiter can give up sufficient angular momentum. This is an interesting idea but it has not been convincingly demonstrated that it will work. An important consideration is the efficiency of the process — as seen in Figure 22.3, the spiral waves are extensive, of the order of many au in extent, while the planets are very small targets for the absorption of angular momentum and energy. For this reason it seems doubtful that they can be greatly affected by this process, especially as there is limited amount of time (a few million years at most) available for the process to operate.

In Chapter 16, the observation of dust disks around older stars was shown to give the possibility, perhaps a strong possibility — no more than that — of planets at much greater distances from stars than Neptune is from the Sun. If their presence were to be positively validated then they would present a challenge to the SNT even more difficult than that of the outer solar-system planets.

The material associated with the inwardly moving spiral waves has less angular momentum per unit mass than the material into which it moves. The effect of this is that as the spiral waves dissipate, they remove angular momentum from the medium.

22.5. Saving the Planet

When a planet is moving inwards towards the star under either Type I or Type II migration, its motion is in the form of a shallow spiral. At any particular time it is moving closely in the form of a circular orbit around the central star. Once it gets close to the star the tidal forces can become large enough to give significant distortion to each body. Here we are just concerned about the distortion of the star.

The mechanism that is to be described depends on two periods, that of the planet in its orbit and that of the star's spin where the latter is to be smaller than the former. To fix our minds we take a star of solar mass with a planet in a circular orbit of radius 0.05 au.

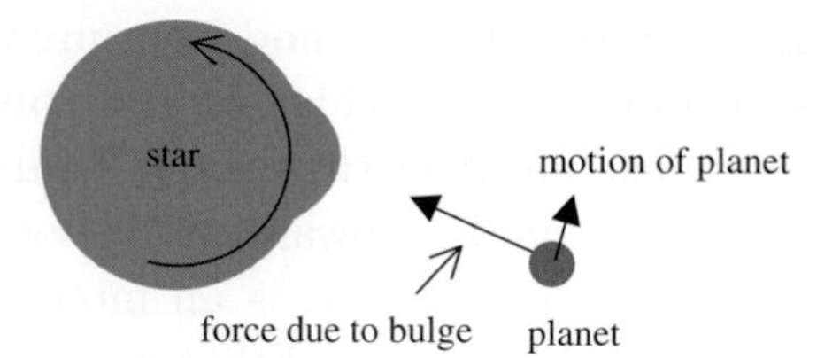

Figure 22.4. Forces on an orbiting planet due to the tidal bulge on a rapidly spinning star.

Such an orbit has a period of about four days and we now take the spin period of the star as three days. Such a star is spinning about nine times faster than the Sun but that is not unreasonable for a young star.

As the planet orbits the star so the tidal bulge on the star will tend to face the planet. However, the faster spin of the star drags the bulge in a forward direction; the situation is illustrated in Figure 22.4.

The force on the planet due to the bulk of the star points towards the centre of the star and neither adds nor subtracts from the angular momentum and energy of the planet's orbit. The force due to the bulge is mostly towards the centre but also has a small component in the direction of the planet's motion. This will add both angular momentum and energy to the motion of the planet and so will tend to move it outwards. It can just balance the loss of angular momentum and energy due to motion in the disk so the planet's orbit is then stabilized and it will go no closer to the star. This is quite a stable situation. If the planet were to move inwards then the tidal effect would dominate and push it out again. Conversely if it happened to move outwards then the medium resistance force would dominate and move it in again. The planet would be saved!

22.6. A Problem with Type-I Migration

The type I migration mechanism is very effective — so effective that it gives timescale problems for the formation of the terrestrial planets. The timescale to grow the Earth at a distance of 1 au from the Sun is greater than the time for the growing object to plunge into the Sun — and that would also be true for the planetesimals that would go into forming the Earth. If the Earth *could* form somehow, then for the tidal mechanism described in the previous section to operate, it would have to be with the Earth very close to the Sun and not where it is now. Various attempts are being made to overcome this problem but they involve increasing the mass of the nebula, or other changes to the standard model, that introduce problems elsewhere.

Chapter 23

Back to Top-Down

There is always room at the top.

Daniel Webster (1782–1852)

23.1. Perceived Problems with the SNT

Although the SNT offers mechanisms for almost all aspects of forming planets around stars, it does seem to be constantly battling with timescale and other problems. Again, the mechanisms that underlie it have been theoretically described but not convincingly demonstrated to work — for example, it is still very doubtful that dust can settle into a plane to enable planetesimals to form — an essential first step. The American cosmogonist, Alan Boss, has therefore returned to a top-down approach but one involving principles not previously considered.

In 2000, Boss numerically modelled the rotation of a viscous disk. Since the speed of material in the disk varies continuously from place to place, there is friction between neighbouring portions of material that creates a rather uneven structure including the presence of filaments of material of density much higher than the local average. A representation of the configuration of the material at a particular time, as given by Boss, is shown in Figure 23.1.

Within the disk there is a condensation, marked with an arrow, that has a mass of about five times that of Jupiter. Unfortunately, it did not persist and disappeared after about sixty years of simulated time. There were a number of assumptions in the model that give cause for comment. The first of these relates to the temperature of the

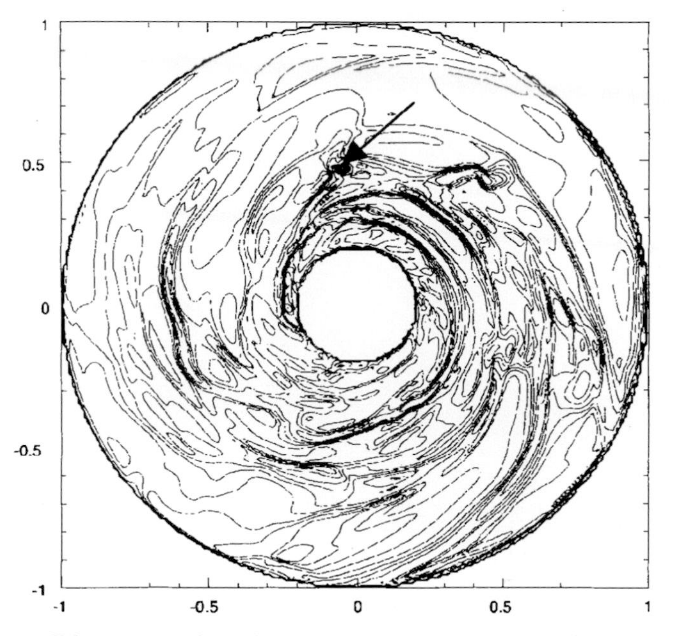

Figure 23.1. Filaments and condensations in a rotating disk of viscous material.

material. Normally, when a gas is compressed, it heats up. Those who have pumped up a bicycle tyre will know that constant compression of the gas makes the pump quite warm. Boss assumed that the gas did not heat up when compressed, an assumption that greatly favours the formation of condensations. However, in later calculations Boss showed that if the material cools quickly when heated then the results are not very different from those without heating.

Another point of doubt is the assumed condition of the disk. It had a mass of 0.091 $M_{\odot}$, just below 0.1 $M_{\odot}$ usually taken as the maximum for a disk, and with radius 20 au, which is the minimum value suggested by Beckwith and Sargent in 1996. This combination of nearly extreme assumptions increases the density in the disk and hence the possibility of producing a condensation.

The new approach is interesting but it still has to prove itself under a range of realistic conditions.

MAKING STARS

Chapter 24

This is the Stuff that Stars are Made of

...an infinite deal of nothing...

William Shakespeare (1564–16516), *The Merchant of Venice*

24.1. The Question

15 g butter
3 eggs
a pinch of salt

Lightly whisk the eggs in a bowl and then melt the butter in a non-stick pan. Strain in the eggs plus the salt and return to a brisk heat. As the mixture cooks at the edge, lift up the edges to allow liquid egg to flow out from underneath. Rock the pan to-and-fro to avoid sticking. When liquid egg ceases to appear from underneath, tip out the contents of the pan onto a plate so that it is neatly folded across the centre.

What we have given above is a recipe for making an omelette. It consists of two parts, the first giving the ingredients and the second describing the process by which the ingredients are transformed into the final product. So it is for the formation of astronomical objects including the stars. Here we shall be mainly concerned with the first part, the ingredients, and we ask the question "What are stars made of?"

A good way to seek an answer to this question is to use powerful telescopes to explore the galaxy and thus to see what possible ingredients there are within it.

24.2. The Galaxy

In the search for stellar ingredients there is no point in looking outside our galaxy since galaxies are well-separated entities, within each of which star formation is taking place. Since we are going to explore the galaxy with our telescope this is a good opportunity to describe the galaxy that we live in — the Milky Way. It has a fairly simple structure. The spiral arms, seen in the similar galaxy NGC 6744 (Figure 2.3), occur within a disk-shaped region. The Sun is situated within the Milky Way disk about one-third of the way out from the centre. Around the centre of the disk there is a spherical region, called the *nucleus* of the galaxy, containing many stars. Finally there is a lightly populated region, the *halo*, in the shape of a flattened sphere, stretching beyond the boundary of the disk (Figure 24.1).

So what do we see in our telescopes when we explore the galaxy? First we see large numbers of individual stars of various kinds, many of them similar to the Sun. Actually some stars are seen to be double stars where the two stars move around each other. There are other, apparently single, stars that can be inferred to be double stars by various kinds of physical measurement — by a wobble in their motions or from Doppler-shifts that indicate that the star we can see has a companion. These double-star systems are called *binary stars* and the interesting conclusion from observations is that the number of binary systems is roughly equal to the number of single stars. Thus only one-third of all stars are single stars like our Sun.

Another type of stellar system revealed by our telescopes is clusters of stars and they are of two main kinds. The first kind normally

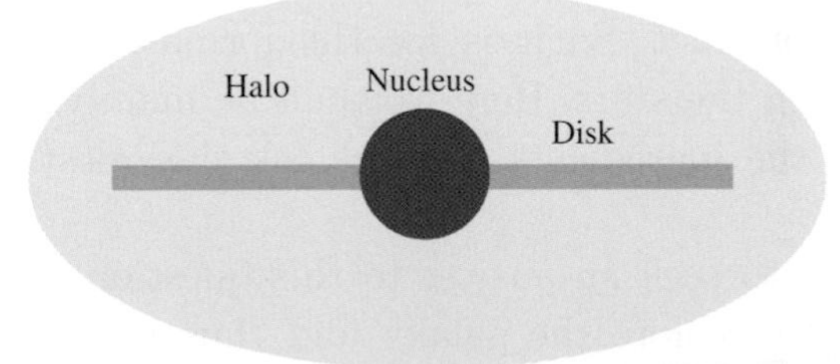

Figure 24.1. A schematic view of the Milky Way galaxy seen edge-on.

contains a few hundred stars that are separated well enough for the individual stars to be easily seen. They also only occur in the disk of the galaxy for which reason they are referred to as *galactic clusters* although, alternatively, in view of their very diffuse nature, they are also referred to as *open clusters*. A typical galactic cluster is shown in Figure 24.2.

The second type of cluster consists of several hundred thousand stars and they are called *globular clusters* (Figure 24.3). They may occur anywhere in the galaxy, including the halo. They are spherical in form and in the central regions, individual stars are not readily resolved.

Figure 24.2. The Pleiades, an open cluster.

Figure 24.3. The globular cluster, M13.

The fact that so many stars are seen in clusters suggests something about the way that stars are formed. Indeed, many astronomers believe that *all* stars, including single stars like the Sun as well as binary stars, are formed in clusters. Individual stars, through gravitational interactions with other stars, may sometimes acquire enough speed to be able to escape from the cluster and calculations suggest that after one hundred million years or so an open cluster will have completely evaporated. Indeed, the term *evaporation* is very appropriate; a pool of water evaporates because individual water molecules happen to acquire enough speed to escape from the surface and eventually the whole pool disappears in this way.

24.3. The Ingredients

But now let us return to the question of ingredients. Clearly the best place to detect the ingredients for making stars is in a region where stars are being made and we can actually locate such regions. One such is the Orion Nebula (Figure 24.4), a region of gas and dust within which newly-formed bright stars and also forming cool stars are observed. This gas and dust is obviously what the stars are being made of and there are several other star-forming regions rich in gas and dust. However, a moments thought leads us to wonder about the

Figure 24.4. The Orion nebula — a star forming region. It is illuminated by four very young bright stars contained within it.

origin of these clouds of dust and gas — are they really the ultimate source, the root source? We swing our telescope hither and thither and we see nothing else. Distant stars are seen, bright and clear and with nothing impeding the passage of light. The space between stars and outside the star-forming clouds is apparently empty.

But wait. Although we can see stars at great distances, when we observe carefully we find that the further away a star is, the redder the light that comes from it. It requires accurate observations to confirm this effect but there can be no doubt about it. From theory, we know that tiny dust particles could have produced such reddening but they must be very thinly spread if the light from stars tens of thousands of light years away can reach us without being blocked out. In fact, these dust particles are part of what is called the interstellar medium, usually referred to as the ISM, which occupies all the space between stars in the galaxy. The dust that causes the reddening of light is only about one percent by mass of the ISM, the rest being gas — mostly hydrogen. Let us create a picture of how diffuse the ISM really is. Look at the last joint of your little finger. That volume in the ISM contains one or two hydrogen atoms. That same volume in the air around you contains about 10^{20} nitrogen and oxygen atoms. The particles of dust are very tiny; about ten thousand of them would fit in the dot above the letter *i*. Now imagine a cube with a side of one kilometre. In such a cube within the ISM there is *one* particle of dust! Actually, the ISM is not completely uniform but is somewhat lumpy. In many regions it will have the composition as just described and a temperature of about 7 000 K. In other regions it may be more or less dense, with respectively somewhat lower or higher temperatures.

Ultimately, this is the stuff that stars are made of, the ISM, but we shall see by what process such ingredients, in such an unlikely form, can be made into stars.

Chapter 25

Making Dense Cool Clouds

. . . and maketh the cloud . . . .

Book of Common Prayer

25.1. The ISM, Clouds and Temperature

It is clear from observations that star formation takes place within clouds of gas and dust that are much denser than the ISM. They are also much cooler and the pattern is that, typically, a dense cloud will be about one thousand times as dense as the ISM with a temperature of about 10 K — a small fraction of the ISM temperature. In the passage from the very diffuse ISM to the production of a dense star (the average density of the Sun is 1.4 times that of water) the formation of a dense cool cloud (DCC) is an essential first step. Here we shall consider a mechanism by which that step can take place.

When we examined the galaxy with a big telescope, as described in Chapter 24, what we could *not* detect was that energy of one sort or another is constantly traversing the space between stars. One obvious energy source is *starlight*, without which we would be unable to see the stars. This is generally very weak but in regions near a very bright star it can be quite significant. The other source of energy is *cosmic rays*. These are not really *rays* in the sense of being a form of radiation but rather they are charged particles moving at speeds very close to the speed of light, and therefore possessing very large energies. They permeate the whole of the galaxy and come from all directions. One of the problems occupying the attention of astronomers is that of the source of cosmic rays but whatever their source, there is no doubt of their existence.

A characteristic of this energy traversing the galaxy is that it is absorbed by matter and when that happens, the matter is heated. This will be so for both the matter that comprises the ISM and also that of the DCCs. Since the ISM and DCCs do not steadily heat up then clearly there must be some cooling process at work that exactly compensates for the heating. One simple form of cooling is through the radiation of heat by the dust particles. This is exactly similar to the loss of heat from a central-heating radiator — the hotter it is the more heat it radiates. However, there are other important modes of cooling, but to discuss these we need to know more about the nature of the material that comprises the ISM and DCCs.

25.2. Atoms, Ions, Molecules and Electrons

ISM and DCC material consists of atoms, molecules, ions and electrons. Virtually all the mass of an *atom* is contained in its nucleus that contains protons, with positive charges, and neutrons with no charge. Around the nucleus there are negatively charged electrons, equal in number to the protons in the nucleus so that the atom as a whole has no charge since positive and negative charges cancel out. The overall size of an atom is typically 10^{-10} m so that one million of them, side by side, would span the dot over a letter *i*. Most of the volume of an atom is occupied by the electrons; if the size of the nucleus is represented by a fist then the electrons are at a distance of several kilometres! A schematic representation of a carbon atom is given in Figure 25.1.

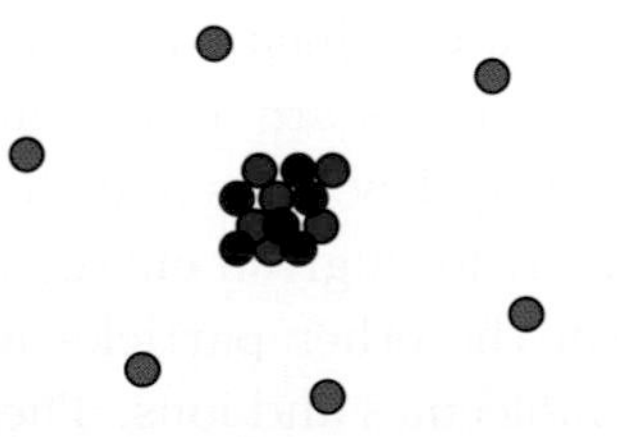

Figure 25.1. A representation of a carbon atom (not to scale). The nucleus contains 6 protons (red) and 6 neutrons (black). The six electrons (blue) exist in a comparatively large region around the nucleus.

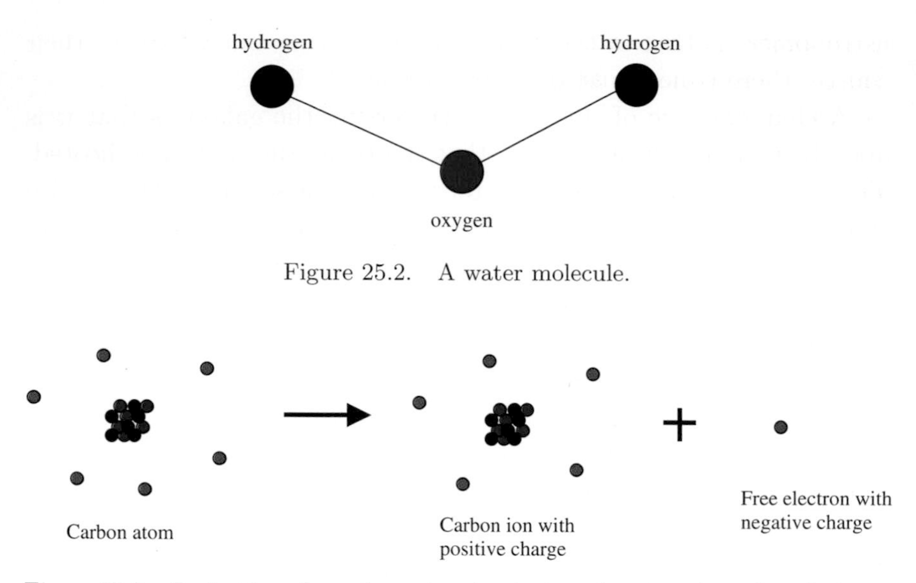

Figure 25.2. A water molecule.

Figure 25.3. Ionization of a carbon atom to give a carbon ion plus a free electron.

Molecules consist of a number of atoms linked together, a well known example being two hydrogen atoms and one oxygen atom joining together to form a molecule of water (Figure 25.2).

Sometimes one or more electrons can be knocked out of an atom. The released electrons, called *free electrons,* can then move around independently leaving behind an *ion* with a net positive charge, as shown in Figure 25.3.

25.3. Further Cooling Processes

As explained in Chapter 2, temperature is a measure of the energy of motion of the particles of which the material consists — for the ISM and DCCs these particles are atoms, molecules, ions and electrons. Because of their very low mass and the equipartition principle described in Section 21.1, for a given energy (temperature) electrons move much faster than the other particles and they are constantly colliding with atoms, molecules and ions. These collisions are responsible for a number of cooling processes but we shall explain just one of them here.

A branch of modern physics, called quantum mechanics, tells us that the electrons in atoms and ions can only exist in states with certain allowed energies. They can jump, or be pushed, from one energy state to another, either up or down, but what they cannot do is exist with an energy that is not one of those allowed. We now consider the collision of a free electron with an atom (or ion), illustrated in Figure 25.4. One of the electrons in the atom can be pushed into an allowed state of higher energy while the free electron correspondingly loses energy of motion. The electron in the atom prefers to return to its original state with a lower energy and it does so. The energy it gives up in this process is converted into a packet of radiation (usually visible or ultra-violet light) and this travels out of the neighbourhood at the speed of light. The net effect is that the original free electron, and consequently the material as a whole, has lost energy of motion and hence reduces in temperature.

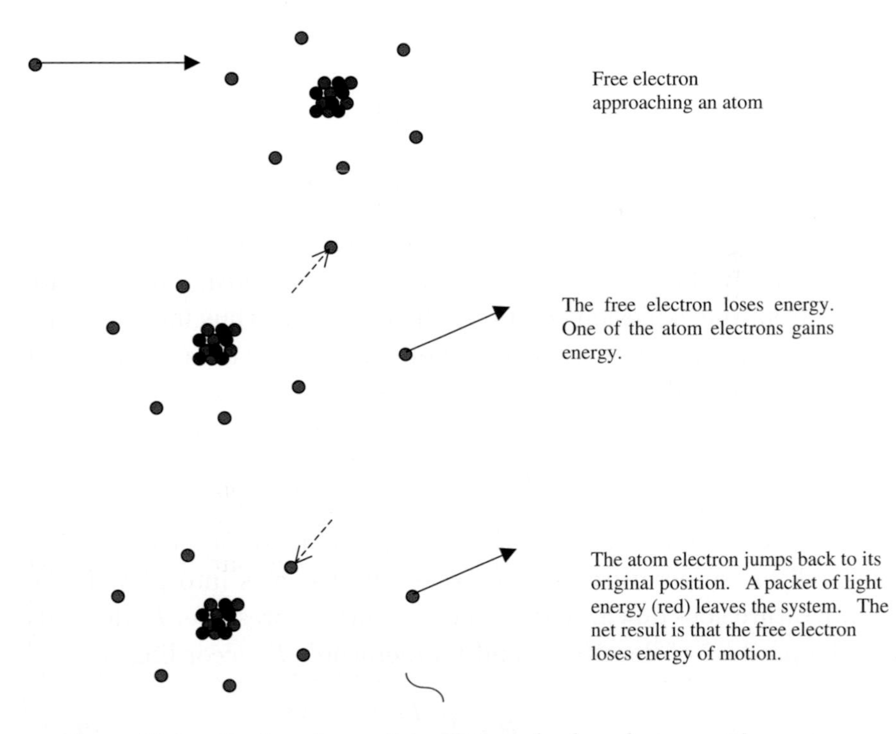

Free electron approaching an atom
The free electron loses energy. One of the atom electrons gains energy.
The atom electron jumps back to its original position. A packet of light energy (red) leaves the system. The net result is that the free electron loses energy of motion.

Figure 25.4. Cooling due to the collision of a free electron with an atom.

There are other processes, some involving molecules, which give cooling. An important feature of all these cooling processes is that they give a greater cooling rate expressed as energy loss per unit mass per unit time when the material density is higher, because with denser material the rate of free-electron, and other collisions, per particle is increased. Another controlling factor is the temperature itself with the obvious relationship that in general, the higher the temperature is the greater is the cooling rate. Knowledge about the various cooling processes that operate in the ISM and DCCs goes back a long way. The British atomic physicist, Michael Seaton, gave the first theory of cooling by the excitation of atomic ions in 1955 and C. Hayashi, a Japanese astrophysicist, analysed the role of dust cooling in 1966.

Although cooling rates depend on both density and temperature, heating rates, especially by cosmic rays, are affected very little by the state of the material and to a first approximation the heating rate can be taken as constant regardless of the form of the material.

25.4. Making a Dense Cool Cloud

To explain the next step in the process of forming stars, illustrated in Figure 25.5, we must assume that some stars already exist. When massive stars get old they reach a stage where they suddenly explode very violently. This kind of event is called a *supernova*. The pressure waves from the explosion compress the local ISM, thus increasing its density, and in addition a great deal of rather dusty debris is injected into the surrounding space. We have seen that when the density of material is increased, the cooling rate also increases and this is what happens to the compressed region of the ISM. Another enhancement of the cooling rate is provided by the extra dust in the region. The heating rate is unchanged but the cooling rate increases, thus leading to a fall in temperature. Now another effect comes into play. If the temperature goes down, so does the pressure — pressure, P, depends on the product of density, ρ, and temperature, T, according to

$$P = \frac{\rho k T}{\mu} \tag{25.1}$$

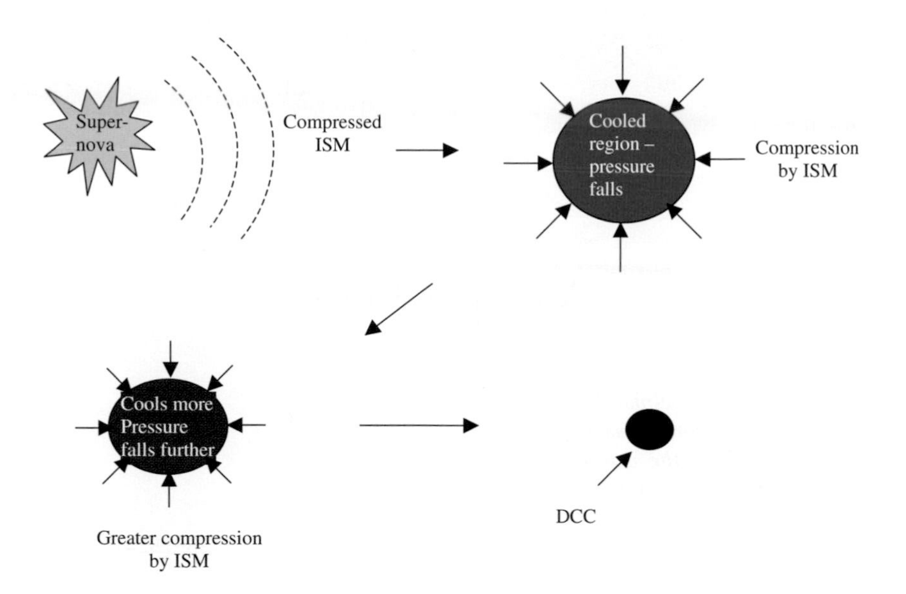

Figure 25.5. The stages in the formation of a DCC, triggered by a supernova.

where k is the Boltzmann constant and μ is the mean molecular mass of the gas. Now the external pressure of the ISM is greater than that within the cooled region so the ISM squeezes the cooled region, compressing it even further and increasing its density. But this further increases the cooling rate and the whole process continues with the density rising and the temperature falling in the cooled region. We are forming a DCC. However, the process eventually comes to an end because although the cooling rate increases with increasing density it also decreases with decreasing temperature and eventually the influence of the low temperature overcomes that of increasing density.

The way that the system adjusts itself is that, at the end of the process, the DCC that has formed is cooling at the same rate as it is heating so that it stays at a constant temperature. In addition, the pressure within it is similar to that of the ISM, so there is no tendency for it either to be compressed by, or to expand into, the ISM. The DCC that has formed is stable, or nearly so, and we have taken the first step towards forming stars. This process has been computationally modelled and was described by Golanski and myself in 2001.

Chapter 26

A Star is Born

Twinkle, twinkle little star,
How I wonder what you are,
Up above the world so high,
Like a diamond in the sky!

Jane Taylor (1783–1824)

26.1. Collapse of Stout Party

We have already noted that when a DCC is formed it is at a steady temperature and is in approximate pressure equilibrium with the ISM. What we did not take into account was the size, and hence the mass, of the cloud. For a typical DCC, with a temperature of 10 K and density 10^{-18} kg m^{-3} (1,000 times that of the ISM) the Jeans critical mass, given by equation (9.2), would be about 1,000 solar masses so that if the DCC mass exceeded that value it would begin to collapse.

A general DCC would have a fairly irregular shape, although the calculation of the critical mass by Jeans was on the basis of a spherical cloud. It is reasonable to take the cloud as spherical for theoretical purposes, as this simplifies calculations about its behaviour, but we have to accept that the numbers we get from our calculations are just indicative of the general value and should not be taken too literally. In this vein, we now consider the way that a spherical DCC with mass greater than the Jeans critical mass would collapse under gravity.

If the cloud material started from rest we might conclude after observing it for a short time that nothing much was happening. What we would be witnessing is what is called *free-fall collapse*. It starts very slowly, gradually gets faster and eventually, in the final stages, is very rapid indeed. Figure 26.1 shows how the radius of the spherical cloud changes with time, slowly at first and then ever more rapidly. Although free-fall collapse does not strictly apply to gaseous material, because it only takes account of gravity forces and ignores pressure effects, it describes reasonably well the initial stages of the collapse of a cloud especially if the mass of the cloud appreciably exceeds the critical mass.

The free-fall time for the total collapse of a uniform sphere was worked out by James Jeans and is given by

$$t_{ff} = \sqrt{\frac{3\pi}{32\rho_0 G}} \tag{26.1}$$

where ρ_0 is the initial density and G the gravitational constant as described for equation (9.1). For a DCC of density 10^{-18} kg m^{-3}, this would be over two million years.

To produce a star of solar mass 2×10^{30} kg, it is necessary to have a density that will make the right-hand side of equation (9.2) at least equal to that value — and preferably greater. For a temperature of, say, 20 K and material with mean molecular mass 4×10^{-27} kg (a mixture of hydrogen, molecular hydrogen and helium) the critical

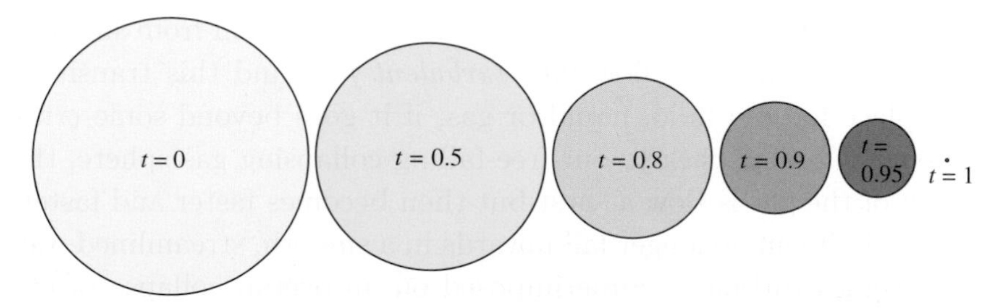

Figure 26.1. The free-fall collapse of a sphere. The times (t) are fractions of the time for complete collapse.

density is just under 10^{-14} kg m^{-3}. This density is 10,000 times greater than the density of the newly-formed cloud, so clearly the cloud, or some part of it, must collapse further before a star can be formed.

When a cloud, or part of it, collapses then it heats up due to compression of the gas. However, the cooling processes, described in Chapter 25, are very efficient and since radiation can pass out of the diffuse cloud quite easily, its temperature remains low. Considering the collapse of the whole cloud, if all its mass fell smoothly towards the centre then we might expect to produce one very massive star containing much of the original mass of the cloud. But that is not what we want so we must now consider the process that instead, produces a large number of normal-mass stars.

26.2. Turbulent Times

In a type of film shown in the early days of the silent cinema one might see the heroine adrift in a canoe on a wide slowly-flowing river. All is well but suddenly the river begins to flow more quickly, although still smoothly. We sense that trouble is ahead. The river is getting narrower and, as it does so, the flow gets faster. Now the river enters a narrow gorge, the flow is not only fast but it is turbulent, with the water violently crashing around. Streams of water travelling in opposite directions crash into one another, spattering water high into the air. The canoe is pummelled this way and that. It is time for our hero to take a hand.

What we have described in this scene is a transition from smooth, also called *streamlined*, flow into *turbulent flow* and this transition takes place for any fluid, liquid or gas, if it goes beyond some critical speed. Coming back to our free-falling collapsing gas sphere, the motion of the gas is slow at first but then becomes faster and faster. Eventually it can no longer fall inwards in a smooth, streamlined way but becomes turbulent. Superimposed on an overall collapse of the sphere there are gas streams moving randomly with respect to each other and occasionally colliding, a situation illustrated schematically

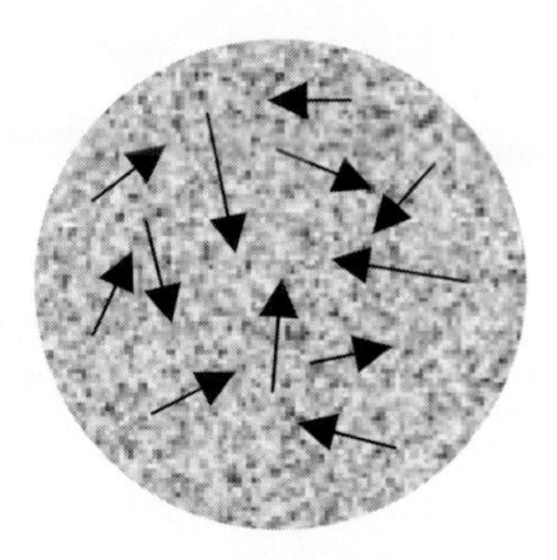

Figure 26.2. A collapsing cloud with turbulence. The general direction of motion of the material (shown by the arrows) is inwards but there is considerable turbulence present.

in Figure 26.2. We shall now see how it is that these colliding streams of gas can give us individual stars.

26.3. The Big Squeeze

When the streams of water collided in the ravine they splattered high into the air. Water cannot be compressed — the streams of water could not go through each other so they just changed direction. Gases are different. When they collide they can be compressed and a clump of gas can form with a density higher than that of the oncoming colliding streams (Figure 26.3).

Here and there in the turbulent collapsing cloud, colliding gas streams produce high-density clumps. The act of compressing the gas also heats it and the mass of the compressed clump will almost certainly be less than the Jeans critical mass for the prevailing density and temperature. However, two things will happen to the clump after it forms. The first is that the gas will begin to re-expand and the second is that the gas will cool, because of the cooling mechanisms we have previously described. The critical factor here is that cooling is a much faster process than re-expansion. Before very long there is a clump of gas slightly less dense than in its original compressed state — but much cooler — and the conditions can then be such that the Jeans critical mass is exceeded. The clump will have a stellar mass and it will begin to collapse. A star is born!

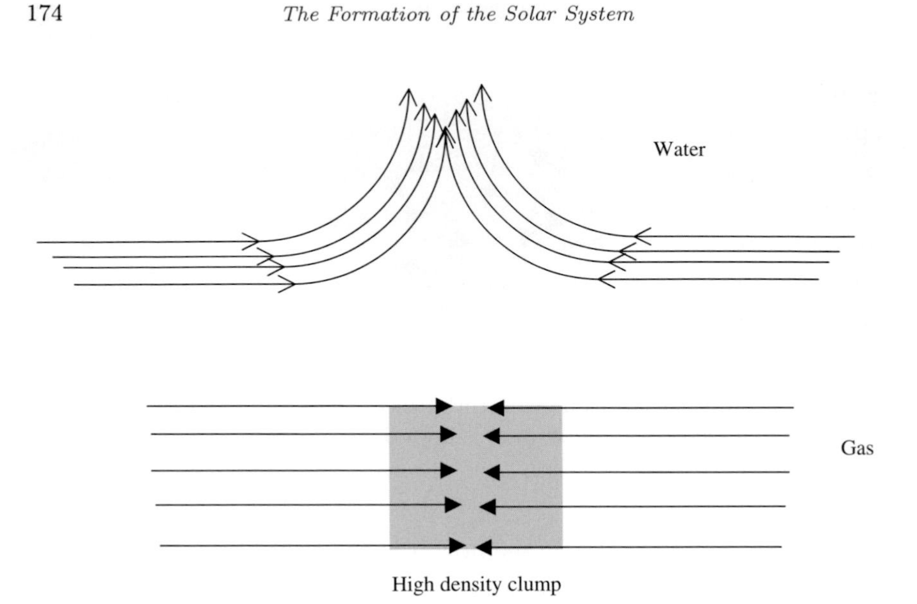

Figure 26.3. The behaviour of streams of water and streams of gas when they collide.

26.4. Some Observations about Star Formation

Turbulence in star-forming regions has actually been observed. Such regions are sources of *maser* (acronym for *microwave amplification by the stimulated emission of radiation*) radiation — like the light from a laser but of much longer wavelength. The source of such radiation at particular characteristic frequencies can be associated with various chemical entities such as water. However, the cause of the maser action that makes the sources so bright is not well understood — but it exists and is easily observable against a low background of microwave radiation covering a wide range of frequency. Observations from star-forming regions show that there are shifts in the characteristic frequencies due to the Doppler effect (Section 15.2) that indicate random radial motions of the source material. These motions are associated with turbulence and turbulent speeds are estimated to be of order 10 km s^{-1}.

For a very young galactic cluster, the new stars are not on the main sequence but are moving towards it on the tracks shown in

Figure 14.3. Having stars with luminosity and temperature corresponding to the region of these tracks both indicate that the cluster is indeed young and also enables the masses and ages of the new stars to be found. In 1969, the British astronomers Iwan Williams and William Cremin examined data for a number of young stellar clusters and were able to deduce certain patterns in the way that stars were formed. Three main conclusions can be drawn from their study, which are, in qualitative form:

(i) The rate at which stars form increases with time.
(ii) The first stars produced are on average somewhat above solar mass and subsequently there are two streams of development, in one of which the average mass of stars increases with time and in the other of which the average mass of stars decreases with time.
(iii) The number of stars produced per unit mass range decreases with increasing mass.

26.5. A Star-Forming Model

In 1979, I investigated a model of a collapsing DCC. The analytical and numerical modelling gave considerable agreement with observations. First the analysis showed that a collapsing cloud would inevitably become turbulent. As the cloud collapsed, gravitational energy was released that turned into energy of motion — some of it in a overall inward motion and some in the form of turbulence, as illustrated in Figure 26.2.

In the modelling of the collision of turbulent elements it had to be taken into account that not every collision would produce a star. First of all, the density of the material must be fairly close to that satisfying the Jeans minimum mass for a star in order that a small amount of compression will raise the density to the right level. If the streams collide too gently then the degree of compression will be insufficient, but conversely, if they collide too violently the material simply gets spattered and will not form a star. The need for a narrow range of collision conditions regulates the rate at which stars are produced and the model gave results closely agreeing with the Williams and Cremin observations — including the feature of an increasing rate of production with time. The first stars produced in the modelling were of 1.3 solar masses and subsequently had less mass; after some time stars that had been produced earlier moved through dense parts of the cloud and were able to accrete material to form a small number of more massive stars. The number of stars per unit mass range found in the numerical results also agreed well with the conclusions from the Williams and Cremin study.

It has been observed that stars with masses less than about 1.3 times that of the Sun tend to spin slowly, like the Sun, while more massive stars have higher rates of spin. The speed of material moving round the equator of the Sun is 2 km s^{-1} whereas for a star with ten times the Sun's mass the equatorial speed would typically be 200 km s^{-1}. The stars produced directly by collisions in my model, all with masses less than 1.3 solar masses, were shown to have low rates of spin — a few times greater than the present values but capable of being brought down to the present values by the mechanism described in Section 18.5. However, for more massive stars produced by further accretion the spin rates from the model were found to be much higher. In fact, the results from the model were in remarkable agreement with observations of spin rates for stars of different masses.

Actually, my original model for the formation of more massive stars, by the accretion of gas onto an existing star, has been thrown into doubt by subsequent work. Ian Bonnell and his colleagues have argued that the radiation from a massive star is so intense that unattached cloud material will be driven away from it rather than being accreted. They have therefore looked at a more plausible model of the coalescence of lower mass protostars to form massive stars. Again, it should be recorded that work announced at the end of 2005 by Mark Krumholz and his colleagues, based on supercomputer numerical modelling, has suggested that once a star forms it cannot accrete much extra mass in the form of unattached cloud material. Actually, the formation of massive stars by the Bonnell process would probably also lead to rapid rotation rates, but this conclusion has not yet been tested.

While there are still many unanswered questions in the matter of the mechanics of star formation there is little dispute that protostars do form and about the way that protostars evolve. When stellar condensations are first produced they are nothing like the stars as we see them in the sky — the Sun, for example. They are large cool spheres of gas — protostars. A protostar with one half the mass of the Sun at a density of 10^{-14} kg m^{-3} has a radius of just under 2,000 au — more than 60 times the radius of Neptune's orbit. Their long journey before they come normal main-sequence stars like the Sun was described in Section 14.2. For stars like the Sun, their hydrogen fuel gives them lifetimes of ten thousand million years. More massive stars have a shorter lifetime, down to a few million years, and when their nuclear fuel is exhausted they cool down and then gravity causes them to collapse in catastrophic events called supernovae. But that is where we began our story about making stars.

CAPTURE

Chapter 27

Close to the Maddening Crowd

Far from the maddening crowd's ignoble strife...

Thomas Gray (1716–1771)

27.1. Neighbours

The Sun is what is called a *field star,* which means that it ploughs its own furrow through the galaxy and is not a member of a cluster or other large association. By measuring the distances of stars in the vicinity of the Sun it is found that, locally, the average distance between stars is about 7 Ly (reminder — a light-year is the distance travelled by light in one year and is about 10^{13} km). The closest star to the Sun is Proxima Centauri at a distance of 4.3 Ly. We found out that the space between the stars was filled by the ISM and it is remarkable that this very diffuse medium, which is somewhat difficult to detect, accounts for almost as much mass as the stars in the solar environment.

We have already mentioned clusters of stars in Chapter 24— open clusters typically containing a few hundred stars and globular clusters containing a few hundred thousand stars. Globular clusters are not really of much interest in our story since the stars they contain are very old, very unlike the Sun. They contain comparatively little in the way of elements heavier than hydrogen and helium, the two lightest elements. Stars like the Sun are found in open clusters and such stars contain one percent or so of heavier elements. It is a common assumption that the Sun began its existence in an open cluster from which it eventually escaped — although there is another possibility, to be mentioned later.

We recognise the existence of a cluster because stars are clumped together to some extent. Clusters vary enormously in the numbers of stars they contain and their dimensions but characteristically, the distance between stars in an open cluster is about 1 Ly. Because clusters are three-dimensional structures this means that the number density of stars (number per unit volume) in a cluster is some 350 ($\sim 7 \times 7 \times 7$) times that in the solar vicinity. This still makes the stars a long distance apart. Taking the Solar System, as defined by the planets, to have a radius of 40 au the distance between stars in an open cluster is more than 800 solar-system diameters.

27.2. Another Big Squeeze

We now return to the collapsing turbulent cloud described in Chapter 26, where stars being produced here and there as streams of gas collide in a suitable way. As the cloud collapses so its density increases and the temperature will also tend to increase although the temperature rise will be greatly moderated by cooling processes. The Jeans critical mass for the cloud material will consequently decrease and stars of smaller mass will tend to be produced as time passes. Of course, the conditions in the cloud will be highly variable; for one thing it will certainly be denser and hotter in its core than in its outer regions. At any time stars with a range of masses will be produced, depending on where they originate, but the average trend will be for stars with smaller masses to form with the passage of time. This is what is deduced from observations. Observations also show that some more massive stars are being produced as the cluster develops and the various ideas about how more massive stars are produced were mentioned in Chapter 26.

The stars being produced in the collapsing cloud are involved in the general inward collapse so that not only the density of the cloud material, but also the number density of stars (i.e. the number of stars per unit volume) within it, would increase with time. Observations of forming stellar clusters in the last few years have shown that they go through an *embedded stage* when the number densities of stars can be extremely high. During the embedded stage, which lasts about five million years, the average distance between stars can be as small as 0.1 Ly or even less. The embedded cluster is held together gravitationally by the large amount of gas it contains, usually with substantially more mass than that of the stars. Eventually the most massive stars go through their lives on the main sequence and undergo supernova events. The energy injected into the cloud will drive out the gas component and the cohesion of the embedded cluster will be lost, thus causing it to expand. In about ten percent of cases what remains at the end is an open cluster. In the other ninety percent of cases the stars completely disperse. Hence, it is more likely that the Sun was part of a cluster that dispersed than that it was a long-time member of an open cluster.

Stars are still produced during the embedded stage so what we have are stellar condensations (that we call *protostars*) with radii of about 2,000 au being formed in an environment where compact stars are about 0.1 Ly (6,500 au) apart. An impression of the relative dimensions of this situation is given in Figure 27.1. The protostars will stay as big diffuse objects for a few thousand years; the free-fall collapse time for a protostar of initial density 10^{-14} kg m^{-3} is, from (26.1), more than 20,000 years.

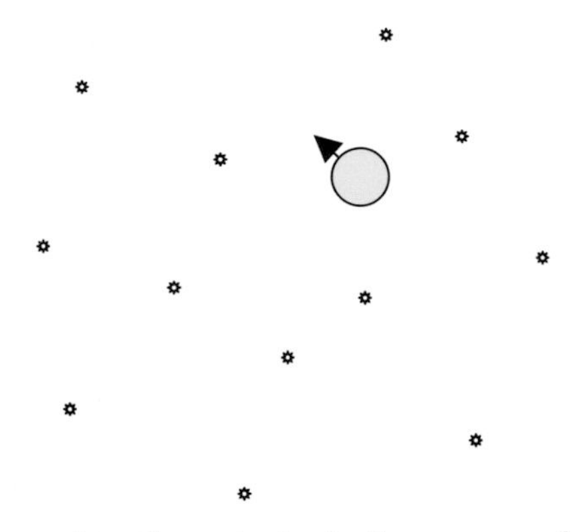

Figure 27.1. An impression of a protostar in the presence of compact stars within an embedded cluster.

In Chapter 26, we mentioned the idea by Bonnell and his colleagues that large mass stars could be produced by the accumulation of numbers of protostars and it is during the embedded stage of a cluster's development that this is most likely to take place. Again, because of the large proportion of stars occurring in binary pairs it is likely that these stars are actually produced as pairs. However, gravitational interactions between stars at the embedded state would tend to disrupt a proportion of the binary systems — a mechanism that has been explored by Kroupa. So, we see that many things can happen in the crowded environment of an embedded cluster.

We now recall the Jeans idea of two tidally interacting stars (Chapter 9). A diffuse protostar would be very greatly affected if it came under the influence of strong tidal forces. So the question arises — "Will an appreciable number of protostars in a diffuse form pass close to condensed stars and, if they do, what will happen to them?"

Chapter 28

Close Encounters of the Stellar Kind

Yet meet we shall, and part...

Samuel Butler (1835–1902)

28.1. Jeans Revisited — The Capture Theory

The two main criticisms that led to the demise of the Jeans tidal theory were as follows:

1 Material from the Sun was too hot to form a planet.
2 Material could not be pulled sufficiently far from the Sun to explain the planetary orbits.

In 1962, I considered an alternative model based on a tidal interaction between the Sun and a diffuse protostar. According to the elegant theory developed by Jeans, the protostar would produce a tidal filament and this filament would break up into a series of blobs that could become protoplanets. The essential difference between this model and that of Jeans is that the protoplanets coming from the protostar were *captured by the Sun* — thus giving the name *Capture Theory* (CT) to the proposed model. Fortunately, this idea arose when computers were improving rapidly and I had access to the most powerful computer in the world at that time, the Ferranti Atlas. Its capabilities would now be regarded as derisory — although it cost over two million pounds it had less than one thousandth of the speed and less than one thousandth of the capacity of a modern desktop machine. Nevertheless, it was good enough to test the capture-theory idea with a simple model and the results were published in 1964. The

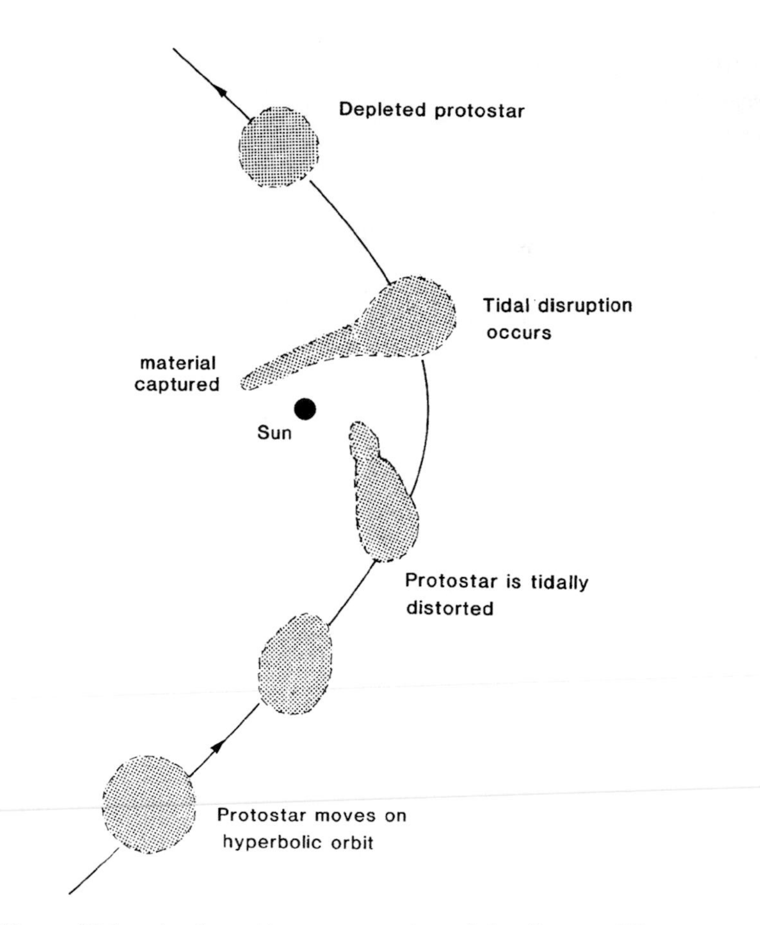

Figure 28.1. A schematic representation of the Capture Theory.

behaviour of the protostar is illustrated in Figure 28.1 in a schematic form that shows what the computation indicated, but is not directly the output from it.

Because of the computer limitations the model was a very simple one, even to the extent of taking the protostar as a two-dimensional object! The protostar was represented as a distribution of mass points, initially placed on a grid, with forces between points representing the effects of gravity, pressure and viscosity. Also present was the gravitational force of the Sun. The scale of this early model was chosen with an eye on the size of the Solar System. The nearest approach of the protostar to the Sun was about 44 au and the radius of the protostar was 14 au. The model was far too crude to show condensations forming in the filament but since Jeans had a good analytical model for this process, various points along the filament were selected as potential condensation centres. Computation showed that material represented by these points went into elliptical orbits that were all within the region around the Sun occupied by the present planets.

This model is free of the objections that had been raised about Jeans' original theory. Since the protoplanet was cool some parts of it could collapse to form planets. Again, since the protoplanet blobs came from the protostar they did not have to be pulled out of, and then move well away from the Sun and the scale of their orbits was mainly determined by the distance of the protostar from the Sun.

28.2. New Knowledge — New Ideas

One of the motivations for creating the model of star formation described in Chapter 26 was to make an estimate of the number of stars that would have planetary systems according to the early model of the CT. A protostar of radius 14 au would have been one that was collapsing quite quickly and therefore the chance of it passing a compact star while it was still in an extended state would have been very small. Another assumption of the 1979 work on star formation was that the separation of stars corresponded to that in a galactic cluster, i.e. with separations about 1 Ly. Assuming the observed relative speed of stars in the solar neighbourhood was 30 km s^{-1} it was concluded that one star in 10^5 would be expected to have planetary companions — which could not be ruled out at that time because the only planetary system known was the Solar System.

The detection of other planetary systems and the realisation that they were reasonably common (Chapter 15) led to a rethink of the conditions of the Capture Theory. The establishment of the existence of an embedded stage in an evolving galactic cluster coupled with the idea of taking protostars in their newly formed state, with radii of order 2 000 au, led to a reformulation of the conditions under which the capture-theory mechanism operated. We shall see that under these conditions, the behaviour of the protostar is somewhat different from that previously found.

28.3. A Method for Realistic Simulations

A very effective technique used in astrophysical simulations is called Smoothed Particle Hydrodynamics (SPH). In SPH, introduced independently in 1977 by Lucy and by Gingold & Monaghan, the astronomical object is represented by a set of points each of which represents some of the object's mass and also carries with it other properties of the material, such as momentum and thermal energy. There are various forces between the points that simulate gravitation, gas pressure and the viscosity of the stellar material. The original 1964 modelling of the CT was, in fact, a very crude form of SPH. During the application of SPH the temperature associated with the points changes due to either compression or expansion of the material and also due to the conversion of kinetic energy of motion into heat energy through the agency of viscosity.

One important piece of physics that was not included in the original 1977 formulation of SPH is that of radiative heat transfer. Heated objects radiate heat energy that is either absorbed by other bodies or lost from the system. Over the years various approximations had been used to take radiative heat transfer into account. At one extreme, if the bodies concerned were very transparent to radiation then it was assumed that they took on the temperature of the local environment — that is to say they stay at the same temperature regardless of expansion, compression or the action of viscosity. At the opposite extreme, if the bodies were extremely opaque to radiation, so that the passage of radiation through them is slow, then it was assumed that temperature changes due to heat transfer by radiation can be ignored by comparison with those due to dynamic effects. Other approximations have been made for intermediate conditions that usually involve taking the temperature of the material as being directly related to its density in some way. However, for this particular application, modelling the CT, a rather better approach was required. Different parts of the model were very different in their characteristics in terms of size and opacity and the central star, irradiating everything in the system, removed any possible relationship between temperature and density. For this reason Stephen Oxley and I devised a new procedure for simulating radiation transfer that closely mimicked the actual physical process that occurs. The whole region of the simulation was divided into smaller regions each of which was a source of radiation, emitted in the form of packets of energy, rather like the packets of energy that physicists call photons, which have already been referred to in Section 25.3. These energy packets were either absorbed by other small regions, so changing their temperature, or they passed out of the system. In the context of radiation transfer the central star was a constant source of radiated energy.

28.4. Capture-Theory Simulations

In the simulation of a CT event there were various parameters that needed to be set. There were:

(i) The mass of the central star (M_*).
(ii) The luminosity of the central star (L_*).
(iii) The mass of the protostar (M_P).
(iv) The radius of the protostar (R_P).
(v) The temperature of the protostar (T_P).
(vi) The composition of the protostar indicated by its mean molecular mass, μ.
(vii) The initial distance of the protostar from the star, D.
(viii) The initial motion of the protostar. This was defined by fixing the closest approach of its centre to the star (q) and the eccentricity of its orbit relative to the star (e).
(ix) The number of SPH points representing the protostar (N).

Figure 28.2 is a simulation, from work by Oxley and myself with the following parameters ($M_\odot$ is the mass of the Sun):

$M_* = 2 \times 10^{30}$ kg $= M_\odot$	$L_* = 4 \times 10^{26}$W	$M_P = 7 \times 10^{29}$ kg $= 0.35M_\odot$
$R_P = 800$ au	$T_P = 20$ K	$\mu = 4 \times 10^{27}$ kg
$D = 1{,}600$ au	$q = 600$ au	$e = 0.95$
$N = 5{,}946$		

The figure shows the SPH points projected onto the plane of the motion. Initially, the protostar is spherical but after 6,000 years, it has approached the star (marked Sun in the figure) and has been greatly distorted. After a further 6,000 years, the whole protostar has been stretched out into a filament and, as would be expected from the work of Jeans, the filament finally breaks up into a series of condensations. These condensations will finally collapse to form planets and we may refer to them as *protoplanets*. Figure 28.3 shows the collapse of a protoplanet from an SPH simulation where the total mass is about five times the mass of Jupiter and in the final stages just over one half of the mass is in the central core while the remainder is in the form of a surrounding disk.

Five of the protoplanets shown in the last image of Figure 28.2 are captured into orbit about the star. The characteristics of these

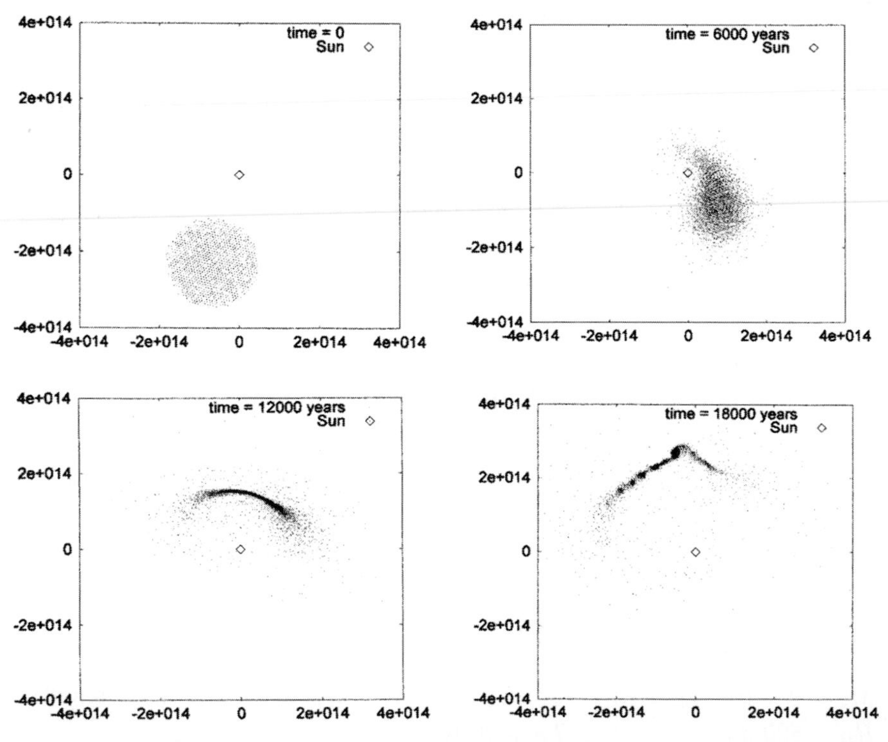

Figure 28.2. A sequence of profiles for a capture-theory SPH simulation. For distances along the axes the terminology 2e + 014 means 2×10^{14}.

captured protoplanets have masses (in Jupiter units), semi-major axes and eccentricities given by

(4.7 M_J, 1,247 au, 0.835), (7.0 M_J, 1,885 au, 0.772),
(4.8 M_J, 1,509 au, 0.765), (6.6 M_J, 1,325 au, 0.726),
(20.5 M_J, 2,686 au, 0.902).

The first four masses are higher than most, but not all, masses observed for exoplanets. However, as we see from Figure 28.3 not all the mass goes into the central core so that the final mass of the planet will be less than that of the original condensation. The final captured body has a mass such that it would not be a planet at all but rather a *brown dwarf*. These are bodies that span the gap between planets and stars. They have the characteristic that they become hot enough to ignite nuclear reactions involving deuterium, an isotope of hydrogen (see Chapter 34) but never hot enough to ignite hydrogen and hence go into a main-sequence state. As a reasonable guide we

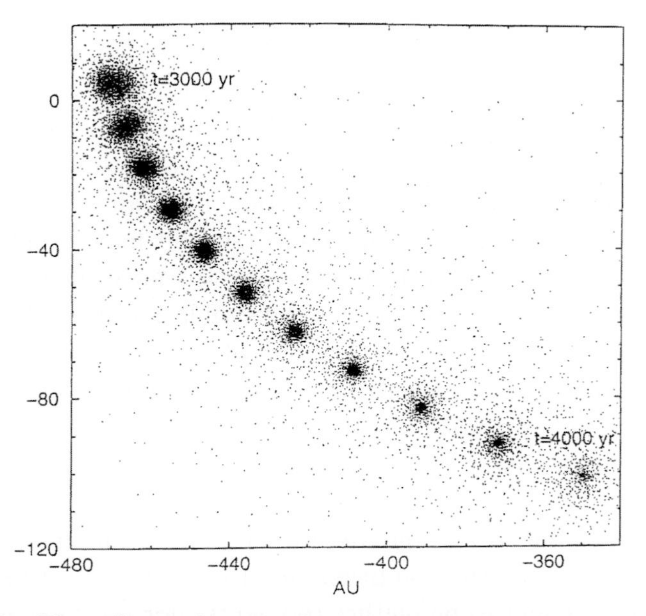

Figure 28.3. The collapse of a protoplanet as it goes into orbit around the Sun, shown at 100 year intervals. At the end of the period it consists of a central collapsing core surrounded by a disk of material.

may take any body with a mass in the range 13–70 times the mass of Jupiter as a brown dwarf. Above that range the body is a star — below that range a planet.

Some of the condensations produced in the filament are of planetary mass but are not captured by the star. They are released into the forming stellar cluster as independent bodies. This feature links up with observations made by Lucas and Roche in 2000. Using infrared observations they were looking for brown dwarfs in the Orion nebula — which they found — but they also found bodies of lower mass that they described as *free-floating planets*.

Another characteristic of the captured protoplanets that needs to be further considered is their orbits. These are much more extensive and eccentric than those that have been observed either in the Solar System or as exoplanets. Indeed there is no possibility that observed orbits could be produced directly by the capture-theory mechanism as described here.

28.5. Doing Without Protostars

A possible problem with the simulation shown in Figure 28.2, and others like it, is that they start with the protostar in the form of a sphere. At the initial distance from the star this is unrealistic. If the protostar formed in that position by a suitable collision of turbulent elements of the cloud then it would have originally have been distorted; if it had formed at greater distance then by the time it reached the starting distance for the simulation it would also have been distorted. Actually, from the point of view of producing planets, starting with a spherical protostar is a *disadvantage*. Since the whole process depends on the protostar being stretched into a filament, starting with a sphere rather than an elongated shape hinders the process.

Work on star formation, described in Section 26.5, indicated that to form a star the conditions for the collision were somewhat restricted — it had to be neither too gentle nor too violent. In fact, in the rather turbulent conditions that prevail in a cloud it is the violent collisions that usually occur and the only way to get a star is if the turbulent elements approached each other rather obliquely so that their mutual relative approach speed was reduced. Oxley and I investigated what would happen to the compressed material produced by fairly violent collisions of streams in the vicinity of a star — something that would happen far more frequently than the interaction of a star with a protostar. In this case, to describe the interaction it was necessary to fix the mass of each stream, their initial positions, velocities, shapes and densities as well as other parameters.

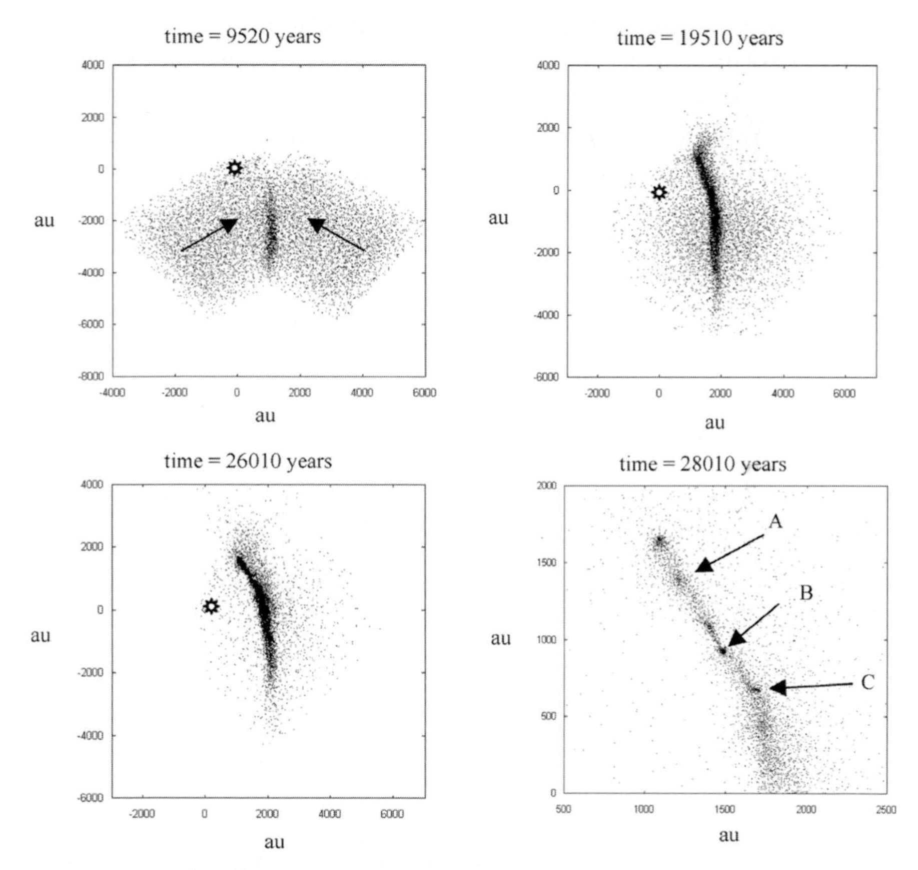

Figure 28.4. A collision between two cylindrical streams of material in the vicinity of a condensed star. The simulation is shown at three times — 9,520 years, 19,510 years and 26,010 years. The final image shows a higher resolution view at 26,010 years.

The result of a collision between two streams, each of mass 0.5 $M_\odot$ and of cylindrical form, is shown in Figure 28.4.

A number of protoplanet condensations are produced, three of which, marked with letters, are captured with mass, semi-major axis and eccentricity as follows:

A (1.00 M_J, 4,867 au, 0.768), B (1.6 M_J, 1,703 au, 0.381), C (0.75 M_J, 1,736 au, 0.818).

In this case condensations with masses less or equal to that of Jupiter are formed but again the problem of modifying the orbits to correspond to what is observed requires attention.

The capture-theory mechanism is extremely robust in the sense that parameters can be varied over very wide ranges to give simulations that produce captured planets. If protostars and high density regions produced by the collision of turbulent elements *were* being produced in an embedded cluster in the presence of compact stars, a scenario supported by observations, then it seems that capture-theory events would inevitably take place. The question is "with what frequency?" and that will be considered in Chapter 30.

Chapter 29

Ever Decreasing Circles

The energies of our system will decay...

A. J. Balfour (1848–1930)

29.1. The Starting Orbits of Planets

The protoplanets produced in the calculations illustrated in Chapter 28 were on orbits very unlike those either of planets in the Solar System or any observed orbits of exoplanets. For one thing the initial orbits of the protoplanets were ellipses with semi-major axes of order 1,000 au and eccentricity 0.9, while for comparison, Neptune, the outermost of the larger solar-system planets, has an almost circular orbit of radius 30 au. The comparison is illustrated in Figure 29.1.

For the CT mechanism of planet production to be plausible, it is necessary to have available some mechanism for producing the required decay and rounding-off of orbits from the initial to the final states.

29.2. A Resisting Medium

The formation of protoplanets, as illustrated in Figures 28.2 and 28.4, not only gave rise to captured and free protoplanets but also to a substantial amount of material captured by the Sun and forming a resisting medium around it. A body, such as a planet, moving in a resisting medium loses energy and this leads to an orbit of ever decreasing size. An example of this, closer to home, is what happens to artificial Earth satellites in low orbits. Although the atmosphere is

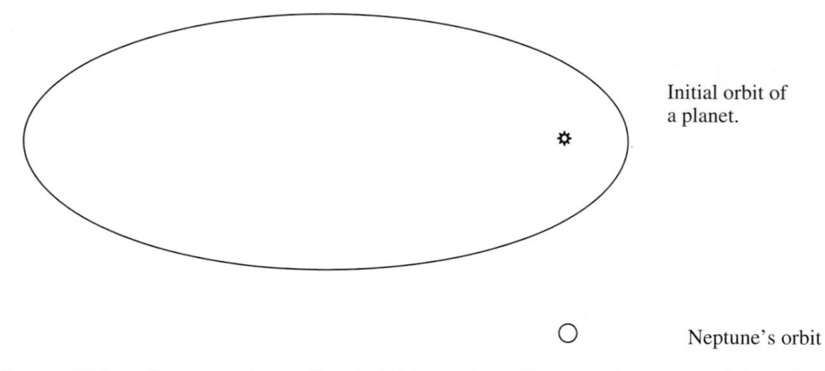

Figure 29.1. A comparison of an initial capture-theory planetary orbit and that of Neptune.

very thin at heights of tens of kilometres above the Earth, it gradually causes the decay of the satellite's orbit, bringing it closer to the Earth. As it approaches the Earth the atmosphere becomes denser and the rate of decay increases. The final stages that lead to the satellite plunging on to the Earth are quite rapid and there is always concern about where the satellite will land — the sea or somewhere in an unpopulated land area being desirable.

The resistance of a fluid to the motion of a solid body within it is a well-known phenomenon in everyday life. Try moving your hand quickly in water — the resistance of the water limits the speed with which you can move it. What is more, the faster you move your hand the greater is the resistance you feel. This is a characteristic of the way that resistance operates. An object that is stationary with respect to a resisting medium experiences no force at all. When it moves with respect to the medium it experiences a force, and the greater the speed of the object relative to the medium the greater is the force. Obviously, the direction of the force is such that it opposes the motion of the object.

Although it is possible to make general statements about the way that resistance operates, it turns out that there are several different mechanisms through which resistance can occur. Of these, one applies to objects of any kind moving in any kind of fluid medium and it depends on the nature of the medium and on the size and shape of the moving object. In another type of mechanism, the mass of the moving object acting on the medium plays a role, and finally there is a mechanism where both the gravitational action of the object on the medium and that of the medium on the object are important.

29.3. Resistance Due to Viscous Drag

It is a matter of everyday observation that water flows easily, that treacle flows less readily and pitch, which becomes fluid on a hot day, flows extremely sluggishly. The property of these materials that leads to their different rates of flow is their *viscosity*, a form of internal friction that inhibits the relative motion of neighbouring fluid layers. If a body moves through a fluid it drags the fluid with it and so causes different parts of the fluid to move with different speeds. The viscosity forces within the fluid react back on the object causing the motion and constitute a resisting force on the moving object.

In Figure 29.2, we see the flow of air around a streamlined object. The flow is smooth and consequently the viscosity force on the object as is small. If the object was of a rather blunt shape, for example a cube, then the flow would be less smooth and there would probably be turbulence, as described in Chapter 26. Turbulence means that neighbouring bits of fluid are moving rapidly with respect to each other and this will generate big viscosity forces and hence high resistance to the motion of the body.

An important characteristic of viscous resistance is that for bodies of similar shape and density it affects more massive bodies less than it does less massive bodies. The reason for this is that the force on

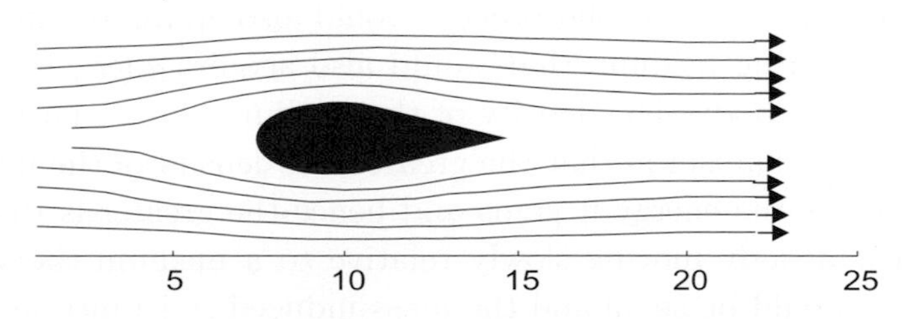

Figure 29.2. Motion of a fluid around a streamlined object.

the body varies with the area impacting in the fluid, which varies as the square of the linear dimension, but this force acts on a mass that is proportional to the cube of the linear dimension. Thus the greater the size of the body the less is the force per unit mass, or deceleration.

29.4. Resistance Due to the Effects of Mass

Let us suppose that the 'resisting medium' through which the body moves does not consist of a fluid but is composed of a large number of more-or-less uniformly spaced solid objects that are so far apart that they do not collide or come close to each other. In this case, there is no viscosity of a conventional kind. The solid objects act on each other through gravity but these forces are just related to the separation of the bodies and not their relative motion.

Now let us imagine that the body is moving through a region occupied by these solid objects, which are initially at rest, without touching any of them. Because of the mass of the body, after it has passed through the region the objects will be moving — i.e. they will possess energy that they did not have originally. This energy has to come from somewhere. Some of it comes from changes in gravitational potential energy due to the rearrangement of the bodies and the rest comes from the body that loses energy by slowing down. This slowdown of its motion is the result of it experiencing a resisting force. This force has nothing to do with viscosity and theory shows that, in many situations, the rate of deceleration of the body is proportional to the mass of the body. What we have deduced for a medium consisting of solid objects would also apply to a fluid, e.g. gaseous, resisting medium that would also acquire energy from the passage of the body. The density of the medium plays a part in this resistance mechanism in that the greater the density of the medium, the greater is the energy it gains and hence the greater is the resistance. For a body moving slowly relative to a medium the viscous resistance would be small and the mass-induced resistance would be dominant.

Finally, for a very dense and extensive medium the gravitational effects of the medium on the body also become important. In this case the passage of the body causes an uneven distribution of the medium due to tidal effects and the clumps of mass so formed will have a direct affect on the body due to their gravitational attractions.

29.5. The Evolution of Planetary Orbits

With the aid of modern computers, it is possible to calculate the way that an orbit evolves with time. In 2003, I carried out such calculations, modelling the medium by a distribution of point masses orbiting the star. While the term 'disk' is usually used to describe what exists around new stars it is not possible for surrounding gaseous material to take on the form of a true disk of uniform thickness. For equilibrium of the various forces acting, the thickness of the disk must increase with increasing distance from the star. Figure 29.3 shows a plan view and a cross section of the disk I used in my model.

The forces due to the medium particles on a model planet in orbit are found and the evolution of the orbit can then be followed. In general, it is found that the orbit both rounds off (becomes more circular) and also decays (becomes smaller). One factor that has to be taken into account is the temporary nature of the resisting medium. This is affected by outward forces due to radiation from the star and also the stellar wind, a stream of charged particles emitted by the star. When stars are young it is very likely that both these forms of emission are much greater than when they reach the main sequence. There is also a process of evaporation whereby the material in the resisting medium gradually escapes into space. We have reasonable estimates of the lifetime of the resisting medium from the inferred lifetimes of disks around new stars — from one to a few million years, at most (Chapter 14).

There is no way of knowing exactly what the initial orbits of the protoplanets were that eventually gave rise to the Solar System. However, it is possible to show by calculations that the orbits can round off from very large eccentric orbits to the sort of orbits we see today. The results of one calculation will illustrate this. A protoplanet

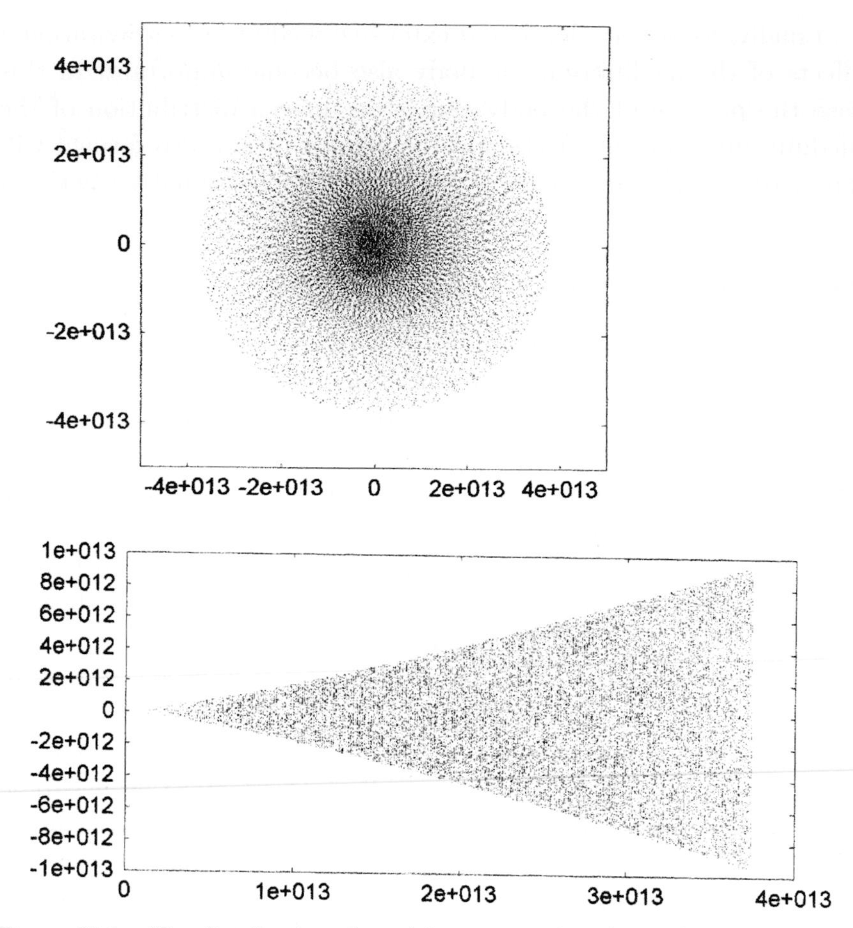

Figure 29.3. The distribution of particles representing the medium in the mean plane of the disk (above) and (below) in a cross section of the disk. (All distances are in metres: for explanation of terminology in representing numbers, see Figure 28.2).

with the mass of Jupiter starts with an orbit with semi-major axis 2,500 au and eccentricity 0.9. To appreciate how big that orbit is, the aphelion is 160 times as far from the Sun as Neptune, the outermost major planet. In the calculation, the resisting medium around the Sun is taken as having a mass fifty times that of Jupiter (0.05 $M_\odot$) with its greatest concentration close to the Sun and gradually falling away with increasing distance from the Sun. The resisting medium also disperses in such a way that it halves its mass every 1.7 million years — i.e. after 1.7 million years one half of it remains and after 3.4 million years one quarter of the original medium is present. After 3.7 million years, the protoplanet has settled down into a perfectly circular orbit with radius 5.3 au, very similar to that of Jupiter. The way that the semi-major axis and eccentricity change with time in this calculation is shown in Figure 29.4. Because the range in the semi-major axis is so large, the vertical plot is done in logarithmic form. This means that changes by a factor of 10, e.g. from 1,000 au to 100 au, from 100 au to 10 au or from 10 au to 1 au all correspond to the same distance along the axis. The behaviour of the medium during the orbital decay is similar to that displayed in Figure 22.3.

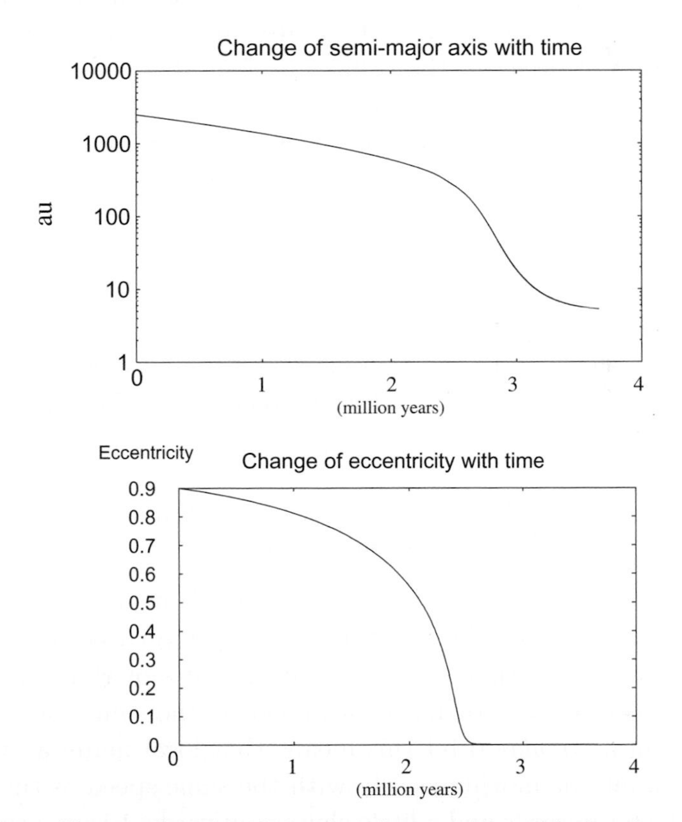

Figure 29.4. Change of semi-major axis and eccentricity with time in a resisting medium.

If several planets are produced by a capture-theory process, as must have happened if the Solar System was produced that way, then each planet would have been affected by the resisting medium and ended up in a more-or-less circular orbit much closer to the star (or Sun) than where it was formed.

Observations of planets around other stars show that some of them are in very close orbits. The total decay of the planet depends on the total mass of the medium, its distribution and its duration. The higher the density of the medium within which the planet moves, the higher will be its rate of decay, and the longer the duration of the medium before it dissipates, the greater will be the total decay. In the calculation leading to Figures 29.4 the mass of the medium was taken as fifty times that of Jupiter but some capture-theory simulations suggest a retained medium with greater mass. That would imply a greater rate of decay of the orbit. On the other hand, observation suggests that some disks decay much more rapidly than that taken in the previous section, which would reduce the total effective time for decay to take place. In several of the decay simulations I found the final semi-major axes less than 0.1 au, as is observed for some exoplanets. The operation of the mechanism described in Section 22.5 could prevent the planets from actually plunging into the star.

Of much greater interest are observed exoplanets with very eccentric orbits (Table 15.1) and to understand these we need to consider in more detail the way in which the forces due to the medium affect the orbit.

29.6. Slowing Down and Speeding Up

The usual assumption of those interested in planet migration (Chapter 22) is that the resisting medium is in free orbit around the central star — that is to say that all parts of it are in circular orbits corresponding to that of a planet at the same distance. For a planet in a *circular* orbit this means that local material, if undisturbed, would be moving closely with the same speed as the planet, a little faster inwards and a little slower outwards. Viscous resistance would be small and the main agencies for orbital change would be the

Type I and Type II migration mechanisms, described in Section 22.3, that are mainly influenced by the mass of the planet and the density of the medium.

For planets in an *elliptic* orbit there are considerable relative speeds of the planet with respect to the medium, particularly at perihelion and aphelion. This gives some viscous resistance and there is further resistance due to the effect of the medium impinging directly on the planet. Whatever the relative importance of the various ways for the medium to apply force on the planet, we can make the general remark that the force is always in a direction that will oppose the motion of the planet relative to the medium, and the force is greater for greater relative speed.

In Figure 29.5, the situation is illustrated for a protoplanet in an elliptic orbit in a freely-rotating medium.

At *periastron* (equivalent position to perihelion for orbit around a star) the planet moves faster than the medium and so is slowed down. The effect of this is to change the orbit to one with less energy (smaller semi-major axis) but the same periastron. This makes the orbit smaller with less eccentricity. At *apastron*, the planet moves slower than the medium so it is speeded up. The effect of this is to change the orbit to one with more energy (larger semi-major axis) but the same apastron. This makes the orbit larger but also with less eccentricity. Hence, at both extremes the effect is to round off

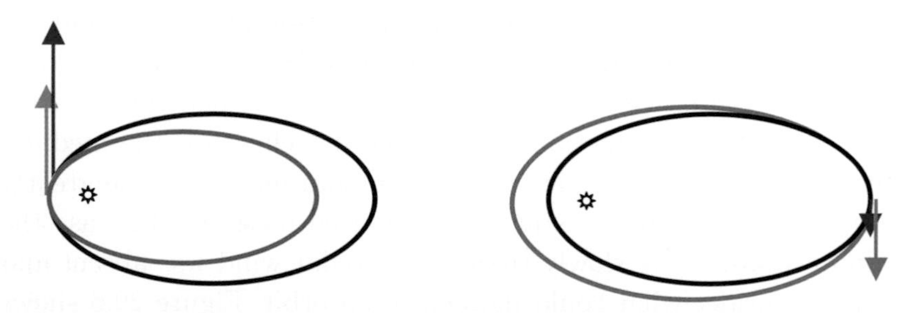

Figure 29.5. Orbital speeds at periastron and apastron marked in red. The medium speed at periastron is shown in blue and the effect of the different speeds slows down the planet and modifies its orbit to the green form. Similarly, the apastron speed of the medium is shown in blue and the effect of different speeds modifies the planets orbit to the blue form.

the orbit but there are opposite effects on the energy, and hence size, of the orbit. If, as is usual, the density of the medium is larger closer in, then the periastron effect will be the stronger and the orbit will round off and decay. Of course, there are resistance forces on the planet at all points on the orbit but the essential features of the orbital modification can be understood just by considering the effects at periastron and apastron.

29.7. Eccentric Orbits

The analysis in the previous section depended on the relative speeds of the planet and the medium at periastron, which in turn assumed that the medium was in free orbit around the star under the sole influence of the star's mass. It has already been mentioned that new stars go through a very active stage where they are more luminous, and have stronger solar winds than when they are on the main-sequence. For example, in 1998 D'Antona and Mazzitelli estimated that the early Sun could have been 60 times as luminous as it is at present and could have had solar winds between ten thousand and one hundred thousand times as strong as it is presently. A strong early stellar wind could have had significant effects on the resisting medium by applying an outward force on it that opposed the gravitational attraction of the star. For the strongest stellar winds, which are believed to be present in young stars in the so-called T-Tauri stage, the wind would completely overcome the gravitational attraction of the star and the resisting medium would be driven outwards. Here we are just going to consider the situation where the stellar wind neutralizes some part of the stellar attraction so that the net effect on the medium (but not on the planet which is too large and massive to be greatly affected) is as though the star had a reduced mass. In this case the medium rotates more slowly than if the stellar wind was absent and we shall consider what could happen to an orbit. Figure 29.6 shows a situation where the medium has been heavily slowed down and the speeds of the planet and the medium at periastron and apastron are indicated.

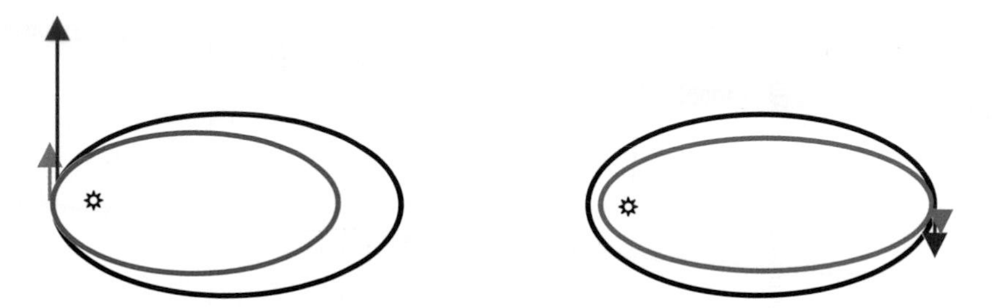

Figure 29.6. Orbital speeds at periastron and apastron marked in red. Medium speed, reduced by the stellar wind, is shown at periastron in blue and the effect of the different speeds modifies the planet orbit to the green form. At apastron, the speed shown in blue is so reduced by the stellar wind that it is now less than the planet speed. Hence, the planet is slowed by the medium and the planet orbit is modified to the blue form.

At periastron, the effect is as previously described — the planet is slowed, the orbit decays and becomes less eccentric. Now, at apastron, the planet is again slowed, the orbit decays and the eccentricity *increases*. The decay is a consistent feature at both extremes of the orbit but the changes in eccentricity oppose each other. In the most common situation the density is higher and the resistance force stronger at periastron so that its effect dominates and the orbit is rounded off. However, in many of the capture-theory simulations that have been carried out, the captured material that forms the resisting medium takes up a doughnut-like form so that when the orbit reaches a certain stage in its development, the medium density is *higher* at apastron than at periastron and it is the effect there that dominates, so that the eccentricity *increases*. This effect was studied in the numerical simulations I carried out and some of the typical results are illustrated in Figure 29.7. The mass of the resisting medium was 50 M_J but a larger mass, which observations would comfortably allow, could considerably shorten the timescales.

In Figure 29.7, simulation K gave a circular final orbit but the other two end up as ellipses, one with an eccentricity of nearly 0.6. By assuming a strong 'doughnut' form to the medium, even higher eccentricities are possible.

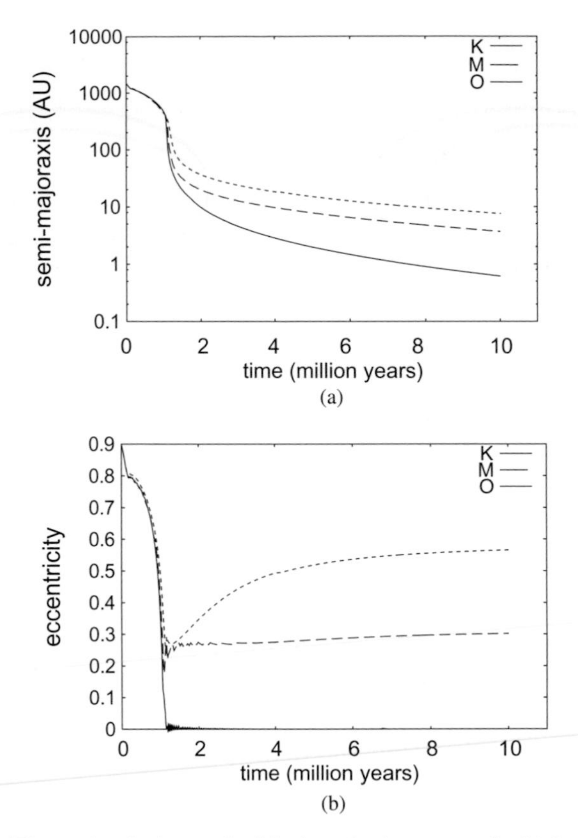

Figure 29.7. Three simulations of orbital evolution, two of which give eccentric orbits.

29.8. Orbital Periods in Simple Ratios

In the process of the rounding-off and decay of the orbits, the protoplanets were not only influenced by the Sun and the resisting medium but also, in some circumstances, by each other. That this might be so is suggested by the fact that the ratio of the orbital periods of pairs of the major planets are very close to the ratio of small integers. For example,

$$\frac{\text{Orbital period of Saturn}}{\text{Orbital period of Jupiter}} = \frac{29.46 \text{ years}}{11.86 \text{ years}} = 2.48 \approx \frac{5}{2}$$

and

$$\frac{\text{Orbital period of Neptune}}{\text{Orbital period of Uranus}} = \frac{164.8 \text{ years}}{84.02 \text{ years}} = 1.96 \approx \frac{2}{1}.$$

In 1996, Melita and I found that there is a mechanism that operates when the orbits of pairs of planets have become circular and are decaying at different rates. When the orbits become *commensurate*, that is when the ratio of their periods equals the ratio of two small integers, then an energy exchange takes place between them. This works in such a way that although the two planetary orbits continue to decay they do so coupled together so that the ratio of their orbital periods remains constant. The effect is a fairly subtle one; while a qualitative theoretical explanation is possible, the mechanism is best explored by calculations. It might be asked why it does not operate between Uranus and Saturn (the ratio of periods here is 2.85). Given that the resisting medium evaporates away then it is possible that it simply did not last long enough for a Uranus : Saturn commensurability to become established — although, given time, the ratio may have become 3.0.

Chapter 30

How Many Planetary Systems?

So many worlds, so much to do...

Alfred, Lord Tennyson (1809–1892)

30.1. More About Embedded Clusters

It is clear that the probability of a star acquiring a family of planets greatly depends on the star number density of the embedded cluster within which it resides — as well as other factors. Before considering this further it is desirable to mention a new unit of length because it is usually used by astronomers to define star number densities (number of stars per unit volume). This unit is the *parsec* (pc), which is 3.09×10^{16} m or 3.26 Ly. It is a unit that arises naturally from a method of measuring the distance of close stars by observing them six months apart against the background of more distant stars, observations taken when the Earth is at opposite ends of an orbital diameter. This was the method suggested originally by Kepler (Section 4.4) but he could not properly exploit it because his measuring instruments were not sufficiently precise. Using this unit of length, the star number density in the solar vicinity can be expressed as $0.08\ \text{pc}^{-3}$, that in a normal galactic cluster about $30\ \text{pc}^{-3}$, that in embedded clusters usually in the range 10^3–$10^4\ \text{pc}^{-3}$ and in the core of the Trapezium cluster, a rich star-forming region, several times 10^4pc^{-3}.

In Chapter 26, the process by which a cloud collapsed, together with the stars in it, to produce the embedded stage was described. The embedded stage then lasts for something in the order of 5 million years, until supernovae from short-lived massive stars drive out the gas, at which stage the cloud re-expands. It is thought that embedded clusters with star number densities in the range 10^3–10^4 pc^{-3} could have had precursor states with densities of 10^5 pc^{-3}, something suggested by Bonnell and colleagues in 2001 — although it also seems possible that the embedded clusters that are observed may be evolving *towards* the state of greatest density rather than away from it.

30.2. Factors to be Considered

We saw in Section 28.4 that planets can be formed by stellar tidal interactions either with a formed protostar or with a high density region produced by the collision of turbulent gas streams. The latter type of interaction will be more common since forming a protostar requires rather special conditions. However, from an analytical point of view, it is simpler to consider the proportion of planetary systems produced by star-protostar interactions. Something *can* be inferred about the number of protostars that are produced, most of which go on to produce stars, but it is very difficult to estimate the number of high-density regions due to turbulent collisions that do *not* produce protostars.

In a moderately dense embedded cluster (star number density 10^4 pc^{-3}), the average distance between stars is just under 10,000 au and the protostars being produced within it have radii from several hundred to about 2,000 au. The speeds of the relative motion of bodies in an embedded cluster have been estimated to be in the range 0.5 to 2 km s^{-1}. The protostar will stay as an extended object for about 10,000 years, as judged by the free-fall time corresponding to its density (Equation 26.1), and for a capture-theory interaction to take place, it must pass close to a compact star within that period. If the simplifying assumption is made that all compact stars have the same mass M^* and all protostars have the same mass M_P then it is possible to determine by computation the proportion of solar-type stars that would be expected to end up with the planetary companions.

It is always important in calculations of this type to be very conservative in the assumptions that are made. The temptation to massage the assumptions in a direction that will give the 'right' result, i.e. that which supports the theory under consideration, must be resisted. Actually, other things being equal, it is always better to put in assumptions that are realistic but in directions that operate *against* the desired outcome for then any positive conclusion will be stronger and more acceptable. For this particular calculation a number of factors must be taken into account. These are:

(i) The distance travelled by a protostar in 10,000 years at 1 km s^{-1} is about 2,000 au. The star will attract the protostar so that its speed relative to the star will increase as it approaches, but even if we ignore this favourable factor, it indicates that the protostar must be formed close to the star if it is to interact while it is still in an extended state. Experience with simulations suggest that a starting distance of between 1.5 and 2.5 of the radius of the protostar ensures that it will be close enough for planet formation, but not so close that it will actually collide with the star. This is a conservative range; in some computations planets have been produced with starting distances both closer and further than given by this range. From the average volume of the cluster occupied by each star and the assumption that the probability of producing a protostar is the same everywhere, the probability that the protostar is formed within the allowed volume for an interaction can be found (Figure 30.1).

(ii) The condition that the orbit of the protostar relative to the star is an ellipse, i.e. has eccentricity less than unity, ensures that at least some of the protostar material will be captured. This condition implies that the speed of the protostar must be less than the escape speed at its point of formation (Figure 30.2). From the average relative speed of stars and protostars within the cluster, the probability that this is so can be found. Again, the condition is conservative. In the capture-theory, simulations carried out by Oxley and myself, and described in Section 28.4, there was one simulation giving captured planets in which the

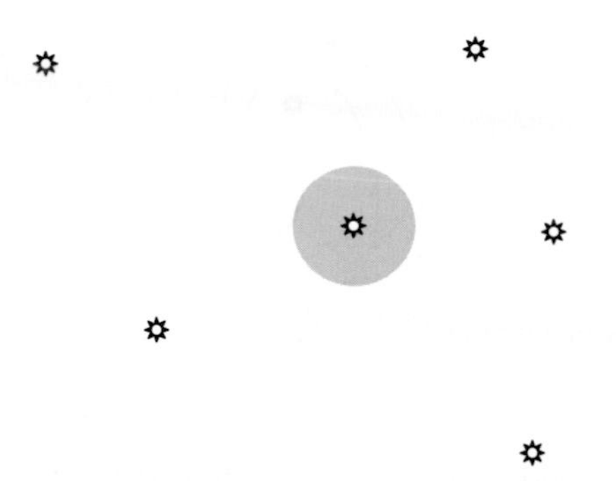

Figure 30.1. A two-dimensional representation of the region for the formation of a protostar that can give a capture process compared with the local distribution of stars.

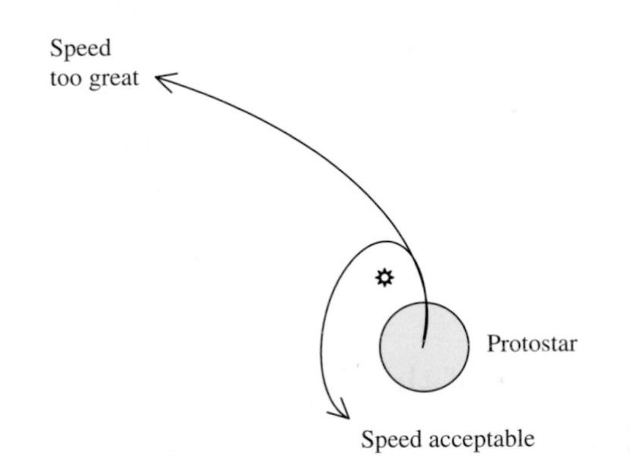

Figure 30.2. An acceptable speed must give an elliptic path.

initial eccentricity of the protostar's orbit was 1.1 — well beyond that for an elliptical orbit.

(iii) For the protostar orbit around the star a condition is imposed that the closest distance of approach should be between $1/2\ R_P$ and R_P where R_P is the radius of the protostar. This ensures an approach close enough to stretch the protostar into a filament. This range would give an actual collision between protostar and star if the protostar did not become distorted, but because of the distortion, no collision takes place. In the simulation described in detail in Section 28.4, the closest approach was about $3/4\ R_P$.

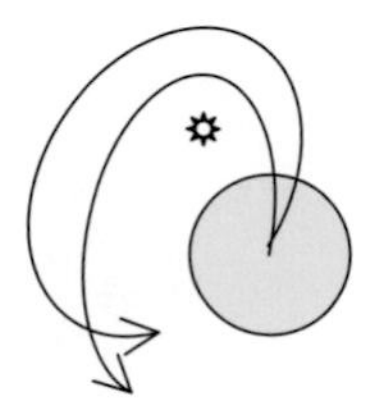

Figure 30.3. For a given speed, the closest approach depends on the direction of motion.

For a given speed satisfying condition (ii) this restricts the initial direction of the protostar (Figure 30.3).

From considerations of this kind, Oxley and I calculated that the proportion of stars expected to possess planetary systems according to the capture-theory model is compatible with the 7% estimate from observations. Indeed, given that stellar interactions with dense regions that are *not* protostars should give a much greater number of planetary systems, the estimated result from theory may seem to be actually too high! However, there is another factor to be taken into account.

30.3. The Ravages of the Embedded Cluster

A newly formed planetary system has planets with extended orbits stretching out from the star up to distances of 3,000 au or more. From time to time, the stars of the cluster approach each other and this raises the question of the stability of the planetary systems. Will they simply be torn apart by passing stars or will a sufficient number of them survive to explain the observed numbers of planetary systems?

On timescales of a million years, the orbits decay down to small size — usually circular, with radius of a few au or less — and once they reach that stage then they are perfectly stable. So it is all a question of timing. If the orbits remained in their original state permanently, then inevitably all the planets would be torn away from the stars; conversely, if the orbits were instantaneously converted to their final state, then none of them would be lost. The true situation lies between these two extremes and the probability of orbital disruption depends on the rate at which the orbit changes, the probabilities of stars approaching each other at various distances and the variation of the star number density with time as the embedded state evolves. Calculations along these lines, carried out by myself in 2004, showed that an appreciable fraction of the planetary systems should survive — although many of them would lose some of the original planetary members of the system. Under a range of conditions that might prevail, anything from one-third to two-thirds of the originally-formed systems would survive.

The capture-theory model does not enable a precise proportion of stars with planetary systems to be estimated. However, the net effect of the estimated number of planetary systems formed plus that of subsequent break up of systems, based on a rather conservative model, gives a result entirely compatible with the present estimate that 7% of solar-type stars have planetary companions — and could even accommodate a higher proportion if future observations so indicate.

Chapter 31

Starting a Family

All happy families resemble one another, but each unhappy family is unhappy in its own way.

Leo Tolstoy (1828–1910), *Anna Karenina*

31.1. The Family Album

Orbiting most of the planets of the Solar System, particularly the larger ones, there are a number of smaller bodies referred to as satellites. Whether these are a general feature of all planetary systems or just a feature of our own system, we have no way of knowing. The answer to that question will depend on the mechanism for forming satellites. If the mechanism is a process that would almost inevitably happen once a planet is formed then we would expect planets in other systems to have satellites. On the other hand, if the mechanism depends on special features of the Solar System, then satellites may not be a general feature of all planetary systems.

To give an idea of the larger members of the satellite community, listed on the following page are all the 15 satellites with diameters greater than 1,000 km, together with their masses and diameters in Moon units and the names of their parent planets.

Satellite	Mass (Moon units)	Diameter (Moon units)	Planet
Ganymede	2.04	1.51	Jupiter
Titan	1.93	1.48	Saturn
Callisto	1.45	1.38	Jupiter
Io	1.21	1.04	Jupiter
Moon	1.00	1.00	Earth
Europa	0.66	0.90	Jupiter
Triton	0.30	0.78	Neptune
Titania	0.048	0.46	Uranus
Oberon	0.039	0.45	Uranus
Rhea	0.034	0.44	Saturn
Iapetus	0.026	0.41	Saturn
Ariel	0.019	0.33	Uranus
Umbriel	0.018	0.34	Uranus
Dione	0.014	0.32	Saturn
Tethys	0.009	0.30	Saturn

In this group of the largest satellites, the Moon is seen as being exceptional in that it is associated with a terrestrial planet, the Earth. Another odd satellite is Triton that, as Regayov noted in Chapter 5, orbits Neptune in a retrograde sense. If these two satellites are removed from the list as particular ones that require some special explanation, then those that remain enable one to make two general statements. The first is that larger satellites are associated with the major planets Jupiter, Saturn and Uranus, and the second is that the more massive is the planet then, in a statistical sense, the more massive are its satellites.

In addition to those listed above, there are a number of satellites with diameters of several hundred kilometres and presumably any explanation for the formation of the larger satellites will also cover these. Finally there are numerous satellites, including both those of Mars, with dimensions of a few kilometres or tens of kilometres, which are best understood as captured bodies. As an example, if two asteroids collide in the vicinity of the planet, then the residual velocity of one or both of them, or of fragments from them, may be lower than the escape speed from the planet. They would then be captured as small satellites. Such an event may also explain Phoebe, the outermost satellite of Saturn, which has a retrograde orbit and a diameter of 220 kilometres.

31.2. The Family Circle (or Disk)

In Figure 28.3, we were shown a simulation of a collapsing proto-planet as it went into orbit around the Sun (or another star). It was in the form of a rapidly collapsing central core with an extensive disk formed around it. The disk had a radius of a few au and it contained about as much material as the core itself. Clearly, it is a good potential source of satellite material so we now have to ask what kinds of mechanisms could produce satellites from it.

One possible mechanism is linked to the Jeans tidal model. A typical disk has a radius of something like 5 au. If the planetary core plus disk, after some orbital evolution, approached the Sun to within 10 au or so, then the disk would be drawn out by tidal effects the form of a filament. In this case, unlike that shown in the simulation in Figure 28.2, the bulk of the mass (that in the core) is not affected by the tide and any condensations — potential satellites — formed in the filament would be left in orbit around the central body (Figure 31.1). None of the objections raised against the Jeans model can be raised against this one. The material is cold and the final distances of planet to satellite are dictated by the extent of the disk and not by how far material can be drawn out of the core. The initial orbits of the satellites would be both highly eccentric and much

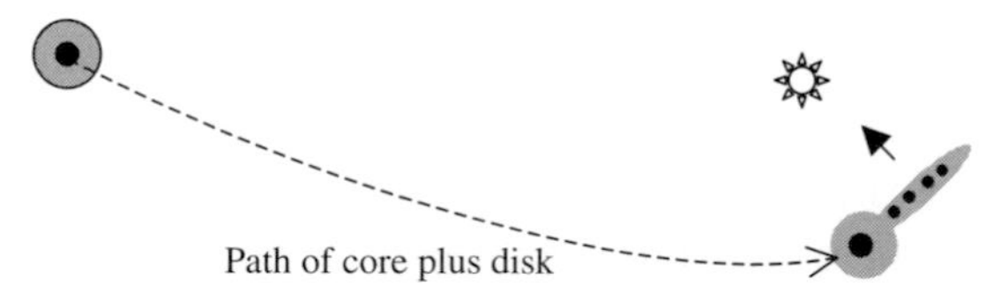

Figure 31.1. Disruption of a circumplanetary disc by tidal interaction with a star.

larger than present satellite orbits but the process of orbital decay, described in Chapter 29, would also apply here at the smaller scale.

Another possible mechanism is along the lines suggested for the formation of planets in the SNT and described in Sections 20.2 to 20.4. In 2004, I considered the formation of satellites using the theory previously developed for the formation of planets by the SNT. Because of the different scale of the system, compared with that for planet formation, timescales are much shorter than in the planetary case. Allowing for particle adhesion, as described by Weidenschilling (Section 20.2) the dust disk settles in a short time. Actually the settling is fastest in the inner part of the disk and I showed that a proposed Jupiter disk would completely settle out to a distance 2×10^{10} m in less than 10,000 years and that the settled region would contain ample solid material to produce the Galilean satellites. The break up of the disk to form *satellitesimals* (analogous to planetesimals) by the process described in Section 20.3 takes only a few years.

The Safronov process by which planetesimals produce terrestrial planets, or planetary cores, as described in Section 20.4 can also occur for satellitesimals. As the protosatellites form, their orbits are constantly changing due to the resisting medium within which they move. The time found using Safronov's formulae, for a satellite to be formed at a distance 2×10^{10} m from the planet was about 10^6 years but is only 40,000 years at 1.6×10^9 m, the orbital radius of Callisto, the outermost Galilean satellite. The actual time of formation of Callisto, which begins to be assembled at a distance of $\sim 2 \times 10^{10}$ m and drifts into a distance of 1.6×10^9 m by the end of the process, is likely to be of the order of a one-to-two hundred thousand years. During this time, the planetary orbit is decaying and by the time the planet is settling close to its final orbit the outer parts of the residue of the satellite-forming disk should either have evaporated or been removed by tidal action with the Sun.

None of the timescale problems that occurred with this model when applied to planet formation occur with satellite formation. One important difference is just that there is ample material in the very centre of the disk, which settles quickly, and that the satellites must move inwards while they are forming — which is an advantage from the timescale point of view.

31.3. Other Features of the Model

We have based the discussion of satellite formation on the Galilean satellites of Jupiter because this is the dominant family within the Solar System. The two outer Galileans, Ganymede and Callisto, have a large ice component while the inner pair, Io and Europa, are rocky — discounting a thin layer of ice covering Europa. A simulation of the collapse of a proto-Jupiter by Schofield and myself in 1982 suggested that a proto-Jupiter would have had a luminosity of 10^{23} W (about 1/4,000 of that of the Sun) during the final stages of its collapse. This would have melted ice up to a distance of 1.3×10^9 m, i.e. somewhere between Ganymede and Callisto. The limit of water ice should be somewhat closer in to accommodate Ganymede's structure, suggesting a lower luminosity of the proto-Jupiter but the general pattern is reasonable.

In Section 29.7, the simple ratios were mentioned connecting the orbital periods of the pairs Jupiter-Saturn and Uranus-Neptune. There are even more remarkable simple ratios for Io, Europa and Ganymede whose orbital periods are closely in the ratios 1:2:4. The coupling of the orbits during orbital decay, as described by Melita and me, is well able to explain this relationship.

Chapter 32

Tilting — But not at Windmills

For sideways would she lean, and sing a faery's song.

John Keats (1795–1821)

32.1. The Leaning Sun

An interesting feature of the Sun is that its spin axis is inclined at 7°. Inclined to what you might ask, and the answer is 7° to a direction at right angles to (or *normal to*) the general plane of the Solar System as defined by the planetary orbits. In fact, the plane of the orbit of the Earth is usually taken as a reference plane for describing planetary orbits and this plane is called the *ecliptic.* The arrangement we have described is illustrated in Figure 32.1.

Now it might be argued that the Sun's spin axis has to point in some direction, and if that direction is at 7° to the normal to the ecliptic, then... so what? That angle is actually quite important. Some theories, e.g. the Solar Nebula Theory (Chapters 18–23) would require the Sun's spin axis to be exactly at right angles to the ecliptic. The actual angle to the normal is small, but large enough for such theories to be concerned and to need to explain the departure from zero. On the other hand, the probability that two random directions in space are 7°or less apart simply by chance is about 1 in 270. This means that any theory giving an initial system in which the plane of the planetary orbits is completely unrelated to the direction of the spin axis of the Sun would need to explain why the angle between the normal and spin axis is so small. The Capture Theory is one such theory since the plane of the planetary orbits is defined by the orbit

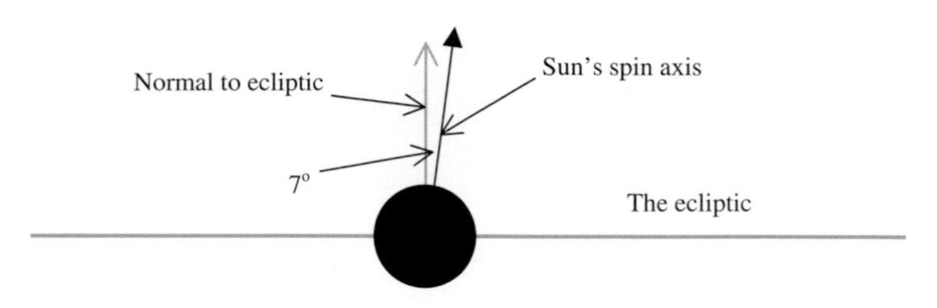

Figure 32.1. The Sun's spin axis in relation to the ecliptic.

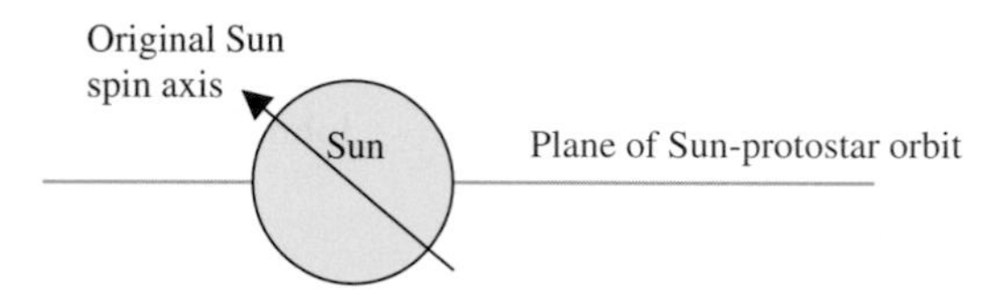

Figure 32.2. A possible arrangement of the plane of the Sun-protostar orbit and the spin axis of the Sun in a capture-theory interaction.

of the protostar (or compressed dense material) around the Sun and has no possible connection with the Sun's spin axis (Figure 32.2). Consequently, we need to explain why the tilt angle for the Sun is so small.

The answer to this problem for the Capture Theory lies in the way that emanations from the Sun — radiation or the solar wind — interact with small solid objects in orbit around it. As already mentioned in Section 29.6, the pressure of these emanations on extremely tiny particles can counterbalance the gravitational field of the Sun, even to the extent that the particles are driven outwards. This effect is negligible for larger particles since the force exerted by the emanations is proportional to the square of the linear dimension of the particle but the mass, on which this force acts, increases as the cube of the linear dimension. Paradoxically, there is another process due to solar radiation falling on larger particles that draws them *inwards* towards the Sun by the mechanism known as the Poynting-Robertson effect (for details see Cole and Woolfson, 2002). For example, a spherical stone sphere of radius 1 cm in the same orbit as the Earth would gradually spiral inwards and join the Sun in about ten million years.

Smaller bodies (but not *very* tiny ones in the form of dust particles) would spiral inwards even more rapidly than that.

We now return to the resisting medium that was described in Chapter 29. Although it was mostly gas, something under 1% of it would be in the form of solid material — tiny grains of ice, stone or iron or combinations of those materials. Some of the grains would have clumped together to form objects large enough to have been sucked inwards and so, in time, to have joined the Sun. The kind of fluffy objects produced by the CODAG experiment (Section 20.2) would be very effectively acted on by the Poynting-Robertson effect. Now these grains orbited in the plane of the planets and the axis around which they orbited was normal to the ecliptic. Consequently, when they joined the Sun they would have had the effect of pulling the Sun's spin axis towards the normal (Figure 32.3). It turns out that surprisingly little material joining the Sun is needed to pull the spin axis close to the normal direction. Depending on the original direction and angular speed of the spin, the absorption of something like one-quarter of a Jupiter mass or so, just 1/4000th of the Sun's mass, is sufficient to bring the axis to within 7° of the normal direction. Such a mass of solid material can be provided by about 25 Jupiter masses of resisting medium, a reasonable fraction of the probable mass of a disk, most of which will evaporate off into space.

The actual rate of spin of the Sun might have been somewhat greater when it was first formed than it is now. Various mechanisms,

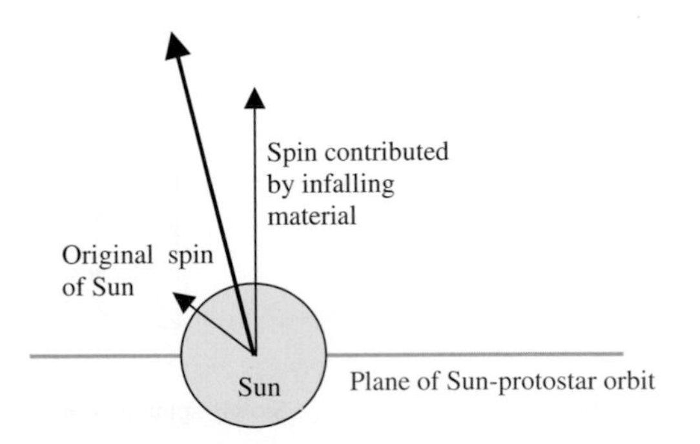

Figure 32.3. The Sun's spin axis is pulled towards the normal to the ecliptic by the addition of material coming from orbits in or near the ecliptic.

especially those involving the way that charged particles emitted by the Sun interact with the Sun's magnetic field can, over time, slow down the spin (e.g. see Section 18.5). A slowdown factor between 2 and 6 over the lifetime of the Sun has been suggested but this would not affect the mechanism we have described for explaining the tilt of the solar spin axis.

32.2. A Child's Top and Evolving Planetary Orbits

There was a time long ago when, at a particular time of the year, most children played with whips and tops. By repeatedly whipping the top, it could be kept in a rapidly-spinning state with the spin axis in a vertical position. However, if the spin rate slowed down, another form of behaviour would ensue. The axis of the top would now be inclined to the vertical and that axis would undergo *precession* round the vertical direction. This is illustrated in Figure 32.4. The reason for the precession of the spin axis is that the force due to gravity acts in such a way as to try to twist the spin axis downwards; the laws of mechanics then operate to give what is seen — although, it must be said that the result that the gravitational force on the spin axis

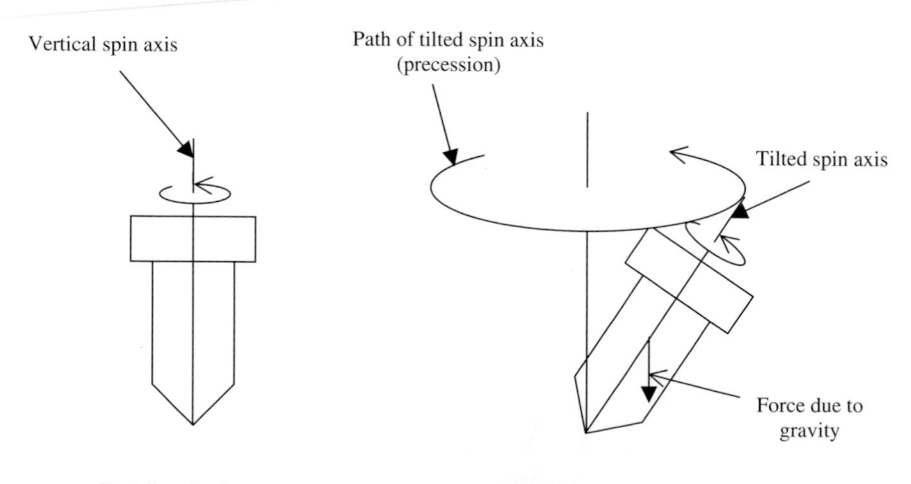

Figure 32.4. A rapidly spinning top spins with a vertical spin axis. When the spin slows, precession of the top occurs.

gives precession comes out of the mechanics of the system and is not at all intuitive.

We now relate the behaviour of the top to the evolution of an elliptical orbit of a planet in the presence of a resisting medium, with the net force on the planet as shown in Figure 32.5. The motion corresponding to the spin of the top is the orbit of the planet. Because the disk of resisting material has mass so the net force on the planet points slightly away from the point about which the orbit takes place — the Sun. This force tries to twist the orbital axis (as gravity did for the spin axis of the top) and hence it causes precession of the spin axis about the normal to the resisting medium. This kind of motion has been confirmed by detailed computer-based calculations.

In the diagram showing precession, the angle made by the orbit with the resisting medium is greatly exaggerated for clarity of presentation. In the capture-theory model the initial orbits of the protoplanets would not be exactly in the plane of the protostar orbit because the spin of the protostar, which could be about an axis in any direction, or the motion of a compressed region, would throw

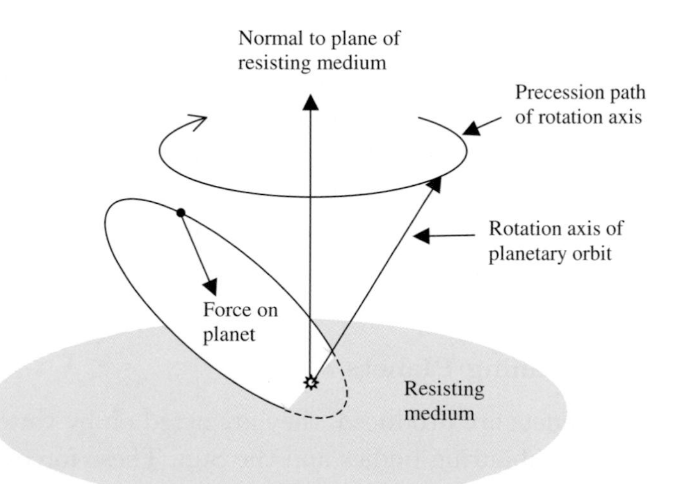

Figure 32.5. The gravitational force of the medium on the planet points slightly away from the Sun and causes precession of the orbit.

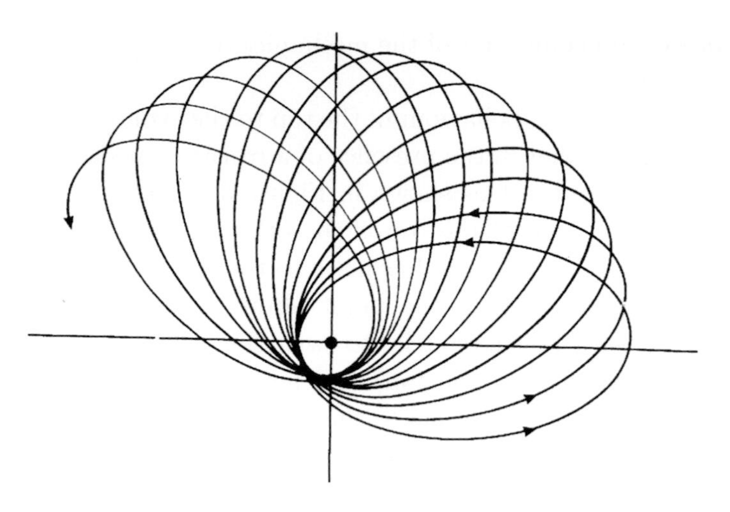

Figure 32.6. The precession of an orbit as seen in plan view.

them slightly out of the plane. However, the angle between the orbit and the plane could be at most about 5°.

In 1977, John Dormand and I showed by computer calculations that the precession of a planetary orbit would be much more complicated than that of a top, with other kinds of motion taking place. From above, projected onto the plane of the resisting medium, the motion would be seen as in Figure 32.6.

The periods for these precessions (the times for the projections of the orbit to return to their original position) are a few hundred thousand years. Since the orbits take about a million years or so to settle down to their final state, this means that there will be a few — typically two or three — complete precession periods during the rounding off and decay process.

32.3. The Leaning Planets

When protoplanets are produced, they are acted on by various forces due to both neighbouring bodies and the Sun. These forces will tend to draw up tides on the protoplanets, rather like the Moon draws up tides on Earth, and this will set up motions of protoplanet material causing them to exhibit spin. Figure 32.7 shows this effect. After the

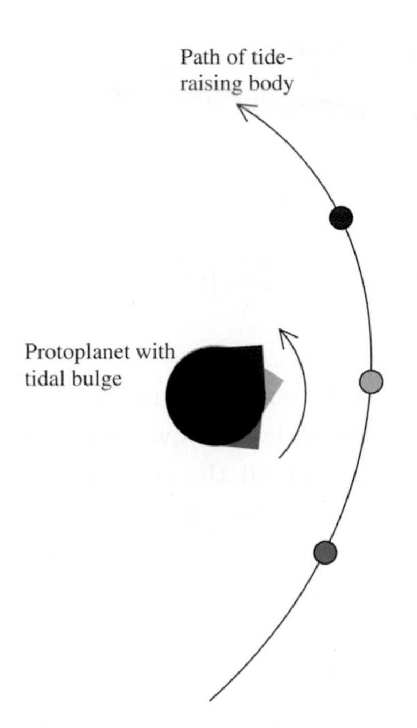

Figure 32.7. The tidal bulge, following the tide-raising body, imparts spin to the protoplanet.

tide-raising body departs, the rotational motion of the tidal bulge spreads throughout the protoplanet giving it an overall spin.

Since the bodies affecting the protoplanet are all close to the plane of the resisting medium, and therefore also close to the plane of the protoplanet's orbit, then it would be expected that the protoplanet's spin axis would not be very far from the normal to its orbital plane. If that were so, then when the protoplanets collapsed to give the planets we see today, that same relationship between the direction of the spin axis and the orbital plane should still be present. However, the pattern for the actual planets is very different, as shown in Figure 32.8.

For all the planets other than Venus and Uranus the sense of the spin is the same as that of the orbit — i.e. anticlockwise as seen from the north. This is indicated by the arrows pointing upwards. For the Earth, Mars, Saturn and Neptune the tilts of the axes are quite marked, from about 23–30°. Mercury has zero tilt and Jupiter's axial

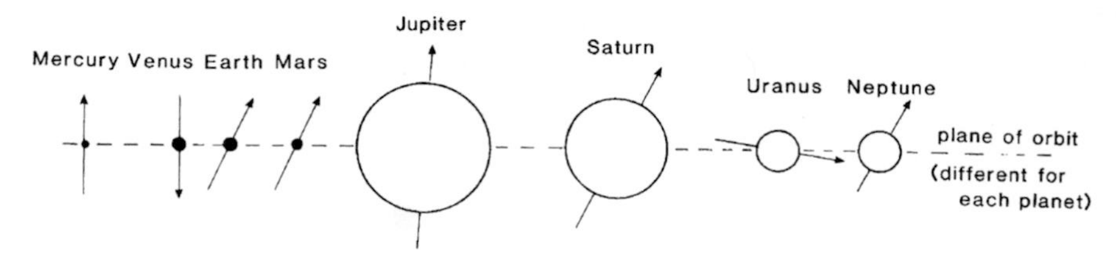

Figure 32.8. The tilts of the planets (excluding pluto) relative to their orbital planes.

tilt is about 3°. The spin axis of Venus is very close to the normal to its orbital plane although its slow spin is *retrograde*, i.e. in the sense opposite from that of all the orbital motions, as seen by the arrow pointing downwards. The tilt of Uranus' spin axis is the most remarkable of all. It is within 8° of the orbital plane of the planet and is retrograde. Seasons on Uranus would be most peculiar. For half its orbit (about 42 years) one pole would point towards the Sun and it would be summer. For the other pole, this period would be an extended winter. For the other half of the orbit the seasons on the poles would be reversed. The greatest challenge is to explain the direction of the spin axis of Uranus; if this can be done, then explanations for other planetary spin axes would surely follow.

32.4. A Fairly Close Encounter of the Protoplanet Kind

If we consider two protoplanets of the Solar System, both on inclined elliptical orbits, then, because their precessions would be at different rates, sometimes their orbits would have intersected, or nearly so, and they could have approached each other closely. Looking down on the orbits, as in Figure 32.9, they may seem to intersect, but if the orbits were slightly inclined to the plane of projection, then the apparent points of intersection would normally correspond to separation of the bodies along the line of sight.

Now let us suppose that a proto-Uranus, with its original spin axis in a direct sense and approximately perpendicular to its orbital plane, passes close to the proto-Jupiter in such a way that the line joining the centres of the two protoplanets is almost perpendicular to both their orbits. Since the proto-Uranus would still be an extended

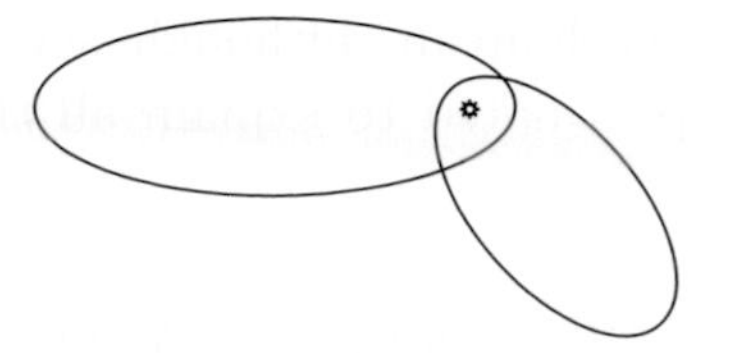

Figure 32.9. Two protoplanet orbits seen in projection.

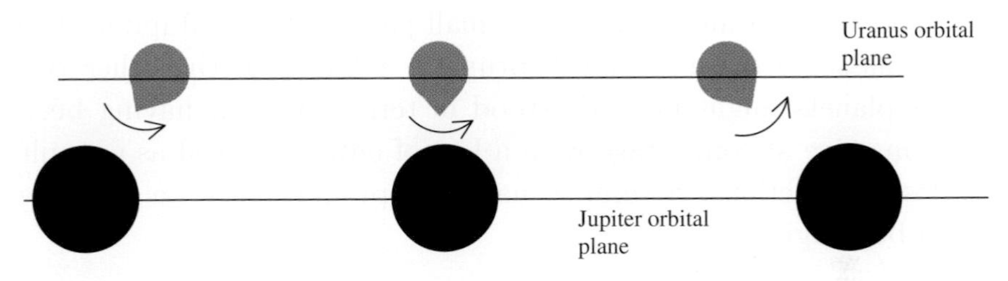

Figure 32.10. The tidal bulge on a proto-Uranus following the passage of a proto-Jupiter.

object (it had not yet collapsed to its final form) the close passage of a massive body would raise huge tides on it and the direction of the tidal bulge would move to point towards the proto-Jupiter as it passed by (Figure 32.10).

The rotation of the tidal-bulge material would be imparted to the remainder of the protoplanet, and in this case lead to a spin axis about an axis almost in the orbital plane. Calculations with model protoplanets have shown that this can take place, and indeed, in one calculation I made in 2000, the tilt of the Uranus spin axis was reproduced almost perfectly. The axial tilts of the other planets can be explained in a similar way — the exact angle of the tilt depending on the geometry of the interaction. As might be expected, since Jupiter is so massive, it is affected relatively little by these tidal processes involving much less massive bodies so that its axial tilt is quite small.

It must be stressed that an exact pattern of the protoplanet interactions that could have given rise to the complete set of planetary axial tilts cannot be determined. We have postulated that it was an interaction with proto-Jupiter that tilted the Uranus axis — but it could have been Saturn, for example. The important thing is that there is a plausible mechanism, although several different sets of interactions may be postulated to explain all the tilts that we now observe.

Another possible result of a close encounter between planets is the loss of some part of its atmosphere. A close interaction between Jupiter and Uranus, for example, which gave such a huge change in the spin direction, might also have removed a large part of the atmosphere of Uranus and even a small part of that of Jupiter. The small masses of Uranus and Neptune in relation to the other two major planets might be understood in terms of them having been more massive at some stage with a loss of outer material as a result of close interactions. Several scenarios to produce such an outcome could be imagined.

THE BIGGISH-BANG HYPOTHESIS

Chapter 33

The Terrestrial Planets Raise Problems!

Notre Père qui êtes aux cieux
Restez-y
Et nous nous resterons sur la terre.

Jaques Prévert (1900–1977)

33.1. The Problem

So far, we have produced a model that gives planets on orbits of the right size accompanied by satellites. Of course, we have not produced a detailed model that produces the Solar System as we see it now, with all planets having the right masses, their proper complement of satellites and in the right orbits. That would be a tall order and it is no mean achievement just to show that the right general characteristics are feasible. Nevertheless there is still a nagging doubt — does this type of model completely give what we should expect?

Capture-theory calculations of the type illustrated in Chapter 28 show that the protoplanets start on extended orbits with their initial motions away from the Sun. By the time they return to the vicinity of the Sun and reach perihelion, most of their material is in a compact central condensation and would be able to resist tidal disruption by the Sun. We have also seen that they would have had surrounding material in the form of an extended disk that would be a source of satellites, probably by the accumulation of solid material. In view of that scenario, one wonders how it is that the terrestrial planets, or small rocky planets, came about. An obvious idea that springs to mind is that since they ended up fairly close to the Sun, they would have heated up and lost their outer volatile material by evaporation so that just a rocky core was left. However, observations of other planetary systems show that large gas planets can exist at distances of 0.04 au from stars that are very similar to the Sun, a distance which is only about one-tenth of the orbital radius of Mercury. On the face of it, the capture-theory model seems likely to be able to produce only large gaseous planets so that some other mechanism for producing the rocky terrestrial planets appears to be required.

Before exploring this question of the origin of the terrestrial planets, we first restrict our attention to the Earth, consider its detailed composition and see what we can learn from that.

33.2. What Kinds of Material does the Universe Contain?

Estimates have been made in various ways of the elemental composition of the cosmos. They cannot be regarded as precise because they are derived indirectly by looking at light from various sources — including the Sun and the Orion nebula, for example — and judging the composition from the strengths of spectral lines corresponding to different kinds of atom. The elements that provide almost all the mass of the cosmos are listed in Table 33.1. The indication seems to be that something like 98.3% of the total mass of the material is hydrogen and helium and heavier elements account for the remaining 1.7%. About 4% of the mass of this remaining material is silicon, the basis of the silicates that comprise most, but not all, of the stony material in the Solar System. Analysis of terrestrial, lunar and meteorite material suggests that the element silicon accounts for about one-fifth of the mass of stony material. That being so, then it seems that about 20% of the mass of the heavier-element material in the cosmos goes into stones of one sort or another. The bulk of the remaining heavier material is mostly oxygen, carbon, nitrogen and inert gases, which exist in the form of gas or ice such as water ice (H_2O), solid carbon dioxide (CO_2), solid ammonia (NH_3) and solid methane (CH_4).

Table 33.1. The main elemental components of the cosmos.

Element	**Relative mass (Si = 30)**
Hydrogen, H	31,800
Helium, He	8840
Oxygen, O	354
Carbon, C	141
Nitrogen, N	51
Neon, Ne	69
Magnesium, Mg	25
Silicon, Si	30
Aluminium, Al	23
Iron, Fe	46
Sulphur, S	16
Calcium, Ca	3
Sodium, Na	1
Potassium, K	2

33.3. What Kinds of Material Does the Earth Contain?

An important component of the Earth, and presumably, of the other terrestrial planets is iron, which exists both in the free form (combined with a little nickel) in the core and as a component of minerals, such as olivine that forms the major part of the Earth's mantle. This division into free iron and mineral iron also occurs in stony meteorites of which the main type is the so-called *chondrites* because they contain *chondrules*, small glassy spheres. Chondrites are further divided into three sub-types.

- *Enstatite chondrites*, in which all the iron occurs either as the metal or as troilite (iron sulphide) and which have a low magnesium : silicon ratio.
- *Carbonaceous chondrites*, where much of the iron occurs as magnetite (Fe_3O_4) and which contain volatile lighter elements and sometimes minerals with water of crystallization.
- *Ordinary chondrites*, by far the most common type which are intermediate in properties between the other two.

Table 33.2. Percentage of total iron and free iron in the three classes of ordinary chondrites.

	H	**L**	**LL**
Total iron (%)	27	23	20
Metallic iron (%)	12–20	5–10	2

In terms of iron content, the ordinary chondrites fall into three classes — H (high iron), L (low iron) and LL (low iron — low metal). The content of these in terms of their iron content is given in Table 33.2.

The usual interpretation of these three classes is that they represent different extents of loss of metallic iron during some high temperature episode in the Solar System, and if that is so, then the H class, with the highest proportion of metallic iron, must be the closest to the original material.

The proposition that we make here is that the Earth is the residue of some large body of material that originally contained the cosmic mix of elements and that the iron within it was divided between the free and mineral forms in the ratio 20 : 7 as suggested by the H class of ordinary chondrites. There are some quite good models of the interior structure of the Earth from which it may be inferred that 35% of the Earth's mass is in the form of free iron in the core and if all the free iron in the original body had been retained, then the total mass of iron, including that in minerals, would originally be

$$M_{Fe} = 0.35 \times \frac{27}{20} M_{\oplus} = 0.473 M_{\oplus}, \tag{33.1}$$

where $M_{\oplus}$ is the mass of the Earth.

From this result we may infer that there is a deficiency of silicon in the Earth. On the basis that 20% of the mass of silicates is silicon, then the 0.65 $M_{\oplus}$ of stony material in the Earth contains 0.13 $M_{\oplus}$ of silicon. However, based on the ratio of the masses of iron and silicon in Table 33.1, the expected amount of silicon from (33.1) is

$$M_{Si} = 0.473 \times \frac{30}{46} M_{\oplus} = 0.308 M_{\oplus}. \tag{33.2}$$

This indicates a loss of more than one half (a fraction 0.178/ 0.308) of the original silicon in the form of silicates.

From Table 33.1, we may also estimate the mass of the original body from which the Earth is assumed to have been derived. Scaling to the amount of iron indicated by (33.1) this mass would have been

$$M_{total} = 0.473 \times \frac{41601}{46} M_{\oplus} = 428 M_{\oplus}, \tag{33.3}$$

which is about 1.35 $M_{Jupiter}$.

The numerical details of these calculations cannot be strongly defended since they depend on a number of assumptions. The estimates of the cosmic abundance of elements varies somewhat with the source, although not widely. The model of the Earth may be in error and some of the core may not be free iron but iron in combination with sulphur and oxygen. Again, it has been *assumed* that the H class of ordinary chondrites can be taken as the model for the composition of the original mass of material of which the Earth is the residue. Nevertheless, even given these uncertainties, it seems very likely that the material of the Earth was once associated with much more material, both in solid and gaseous form.

We now have a final question to answer. If the Earth was originally part of the solid component of a body of greater than Jupiter mass, then how was it derived from that body?

Chapter 34

A Biggish Bang Theory: The Earth and Venus

Thou stand'st unshook amidst a bursting world.

Alexander Pope (1688–1744)

34.1. A Very Close Encounter of a Planetary Kind

In Chapter 32, we saw that the original orbits of the protoplanets were highly eccentric, inclined at a few degrees to the ecliptic and that they underwent precession. It was also shown how near passages of pairs of major planets could give rise to the present tilts of their spin axes. Here we look at an even closer interaction — in fact, a collision.

If the eccentric, inclined and extended orbits of two planets have different precession rates so that their relative configuration continually changes, then from purely geometric considerations, if their orbits intersect as seen in projection (Figure 34.1) then from time to time their orbits will actually intersect in space.

The intersection of orbits does not guarantee that a collision will take place. The other requirement is that the planets should actually reach the point of intersection of the orbits at the same time. Given the characteristics of two planetary orbits, it is possible to calculate how long it would take on average for the two planets to collide. Examples of such calculations were given by Dormand and me in 1977. The average calculated times for pairs of planets to collide turn out to be somewhat greater than the round-off times for the planets; of course, once the planets have rounded off then no collisions can

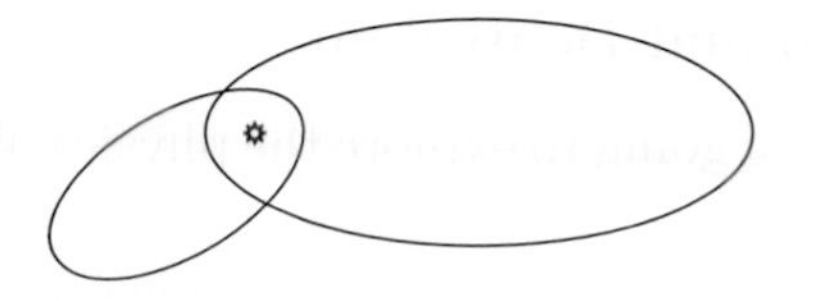

Figure 34.1. Planetary orbits, slightly inclined to the plane of representation, intersecting in projection.

take place. However, it is only the *calculated average* time that is greater than the round-off times and in any real physical situation the *actual* time for a collision may be less. The calculated results suggested that the probability of a planetary collision in the early Solar System may be reasonably large, perhaps 0.1–0.2 or so just to put a figure to it.

As an event, a planetary collision lies well outside the range of everyday experience in terms of the energy produced. Yes, we see things colliding on Earth, sometimes due to tragic accidents, and we have seen films of the most energetic man-made phenomena, thermonuclear explosions. We are told that an asteroid colliding with the Earth some 65 million years ago brought about the extinction of the dinosaurs — and the energy released in such an event would have dwarfed even a thermonuclear explosion. However, the asteroid was probably no more than a few kilometres in diameter, a trivial object compared to a major planet, and even the Earth is small compared with, say, Uranus, the least massive of the major planets. We can carry out calculations to indicate what would have been the outcome of two major planets colliding, but it is difficult to *imagine* such an event. The energy released in the collision would have been at least 10^{18} (a million million million) times as much as in the most powerful thermonuclear explosion. Another comparison is that it would release in one event, of duration measured in hours, the total energy output of the Sun over a three-year period.

Before modelling what would have happened in a collision between two planets, we must first consider some aspects of the material comprising the planets.

34.2. Hydrogen and Deuterium

In Chapter 40, we are going to explore the physics of *isotopes*, different forms of the same element. The simplest of all elements is hydrogen and in its most common form it consists of a single positively-charged proton plus one negatively-charged electron (Figure 34.2a). This atom is represented by the symbol H. There are two other commonly-occurring forms of hydrogen — deuterium (Figure 34.2b), represented as D and containing an uncharged neutron in the nucleus as well as the proton, and tritium (Figure 34.2c), represented as T, that has two neutrons with the proton in the nucleus,

Tritium is not of much interest to us since it does not occur naturally. It does have important medical uses and can be produced in a nuclear reactor. It has a half-life of 12.6 years, which means that at the end of every 12.6 year period, there is only half as much remaining tritium as there was at the beginning. However, the ratios of D/H *are* of interest since both hydrogen and deuterium are products of the 'big bang' that is thought to have created the universe. This ratio has been determined for many solar-system bodies and some of these ratios are

Jupiter (atmosphere)	0.00002
Uranus (atmosphere)	0.00006
Earth	0.00016
Meteorites	up to several times the Earth value
Comets	twice the Earth value
Interplanetary dust particles (IDP)	0.01
Venus	0.02

The Jupiter value is taken as indicative of what was in the original material that produced the planets. Jupiter exerts such a massive

H D T

Figure 34.2 (a) Hydrogen (b) Deuterium (c) Tritium (• = proton, • = neutron, • = electron)

gravitational pull on its own material that it retains everything it ever had. At the other end of the list, we have Venus with a massive 2% of deuterium within its hydrogen. Venus also has the distinction of being a very arid planet having lost most of its water at some stage. The loss of water and the high D/H ratio can be linked, the process being illustrated in Figure 34.3.

We start with Venus with high water content and a water rich atmosphere. Due to its proximity to the Sun plus the greenhouse effect (Section 36.1) its surface temperature was high. Water has the chemical composition H_2O, so that a molecule of water contains two hydrogen atoms plus an atom of oxygen. However, since there is deuterium with the hydrogen, some water will be in the form HDO, and there may even be a little D_2O.

The left-hand side of Figure 34.3 shows one molecule each of H_2O and DHO in the atmosphere of Venus being irradiated by the Sun. The energy of the radiation breaks up H_2O into OH and H while DHO is split up into OH and D. At any given temperature, the average speed of an atom (or molecule) is given by

$$v_{\text{average}} = \sqrt{\frac{3kT}{2m}}, \tag{34.1}$$

where T is the absolute temperature, m is the mass of the atom or molecule and k is a constant — Boltzmann's constant. From this,

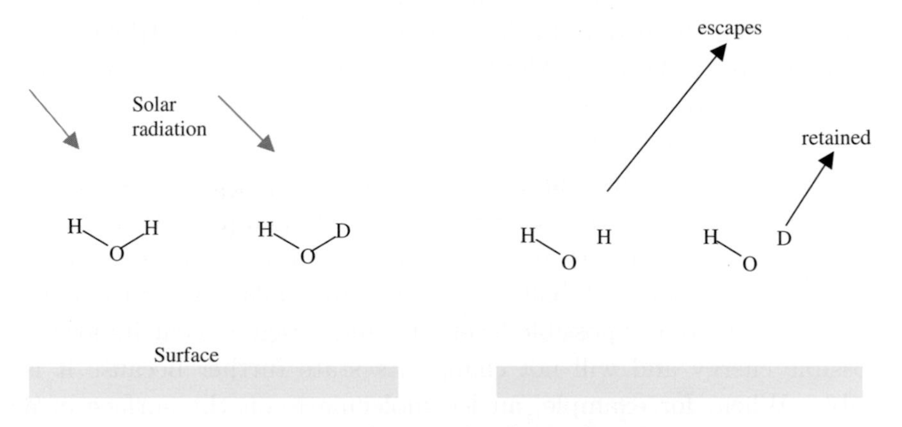

Figure 34.3. The loss of hydrogen from Venus.

we can see that since D has twice the mass of H, it moves more slowly than H and can be retained by Venus while H has sufficient speed to escape. This is only part of the process. Two OH entities that are unstable will combine to form H_2O plus O and the oxygen will combine with other materials that are around. However, the main effect of what we have described is that water is being lost and that the residual hydrogen is becoming enriched in deuterium. We can envisage that some similar process, but to a lesser extent, took place on Earth to give a D/H ratio eight times higher than that on Jupiter.

34.3. Deuterium in Early Planets

While the overall ratio of D/H in the universe is about 2×10^{-5} as is found in Jupiter, the distribution of deuterium is by no means uniform even outside the Solar System. Measurements of spectral lines in the far ultraviolet part of the spectrum enable estimates to be made of the D/H ratio from various sources. Thus in the dense cool cloud L134N the ratio of doubly-deuterated ammonia to normal ammonia, NHD_2/NH_3, is 0.05 and in the low-mass protostar 16293E the ratio is 0.03. The ratio of fully-deuterated to normal formaldehyde, D_2CO/H_2CO, has been found to be in the range 0.01 to 0.4 in a number of low-mass protostars. In the protostar IRAS 16293, the amount of deuterated methanol actually exceeds in amount that containing normal hydrogen. In the same protostar a ratio of HDO/H_2O of 0.03 has been observed. All these observations, and others, indicate that in the cold icy grains that are present in an early protostar, or protoplanet, a high concentration of deuterium should be expected.

These observations can be explained by what happens on the surfaces of ice grains in the presence of hydrogen gas. All physical and chemical systems try to achieve a state of lowest energy where they are most stable. A ball placed on a hill rolls downwards until it reaches the lowest possible level, at which stage it is at its lowest possible energy and will not change its state further because it is stable. When, for example, an ice molecule is on the surface of a grain situated in hydrogen gas then exchanging a hydrogen atom from the molecule with a deuterium atom from the gas lowers the energy state of the molecule. Hence this exchange, which brings about a lower energy state and greater stability, tends to occur and in this way deuterium is gradually concentrated in the ice.

34.4. How to Make a Hydrogen Bomb

One of the least attractive advances in technology of the atomic age is the development of the hydrogen bomb. A more attractive potential technology, related to the hydrogen bomb, is the peaceful generation of electricity by fusion, for which hydrogen is the basic fuel. That would be a comparatively clean source of energy with little in the way of radioactive waste products and without the production of greenhouse gases. In Section 14.2, it was mentioned that in the interior of a star, like the Sun, the temperature is about 15 million degrees and at that temperature the process of converting hydrogen into helium by nuclear reactions takes place. One of the things that may not be realized is how *small* the rate of energy production is in the Sun in relation to its mass. The Sun has a mass of 2×10^{30} kg and an energy output of 4×10^{26} W — i.e. 2×10^{-4} W kg^{-1} being produced by thermonuclear processes. We can compare that with the heat production by an average human who, by chemical processes within the body, produces about 50 W from a body mass of, say, 70 kg — that is 0.7 W kg^{-1}, some 3,500 times the solar value. It is fortunate for us that the Sun does produce energy at the rate it does. If the rate matched that of heat production by a human, then firstly, life would be impossible on Earth and secondly, the Sun will have exhausted its fuel long ago. To reproduce the process taking place in the Sun, using normal hydrogen, at a rate that gives a useful energy output on Earth, would require the production of unachievable densities and temperatures.

The use of a mixture of deuterium and tritium as a fuel makes the process far more attainable and modern fusion research is concerned with obtaining energy from these isotopes. The material must be confined for at least one second at a temperature of 100 million K in order to obtain an output of more energy than is put in; achieving this presents formidable engineering challenges. This mixture also forms the fuel for a hydrogen bomb. It is inserted within a normal atomic bomb, based on uranium or plutonium fission, and when this explodes the conditions for fusion to take place in the hydrogen isotopes is attained.

Another way to obtain the processes that occur in a hydrogen bomb is to get two large planets to collide, especially if they contain deuterium-rich material. In this case, the temperature produced by the collision is typically only a few million K but the density of compressed material can be very high, which increases reaction rates, and the ignition temperature can be maintained over a long period of time. More will be said about this process in Chapter 40. For now, we may just note that when planets collide, thermonuclear energy is produced which adds to the dynamical energy and influences how the collision progresses.

34.5. The Colliding Planets

Something that is deduced from the detection of exoplanets, as illustrated in Table 15.1, is that planets of well above Jupiter mass are quite common. It turns out that if at least one of the colliding planets is considerably more massive than Jupiter, then many quite different features of the Solar System can be explained as by-products of the collision. There is no way of knowing what the actual planetary masses should be so all one can do is to show that with some postulated masses, outcomes are obtained that agree with, and explain, features of the Solar System — although other pairs of masses could also give acceptable outcomes.

The two model planets used to study the collision have masses of 618 $M_\oplus$ and 116 $M_\oplus$, where $M_\oplus$ is the mass of the Earth. In terms of the major planets we know, these are about twice the mass of Jupiter and one-and-a-quarter times the mass of Saturn respectively. These planets were additional to the present four major planets of the Solar System. For the sake of identification, the colliding planets will be named Bellona and Discordia. The mythical character

Bellona was the Roman goddess of war, the sister, wife or daughter of Mars — mythology seems uncertain about the relationship. Discordia was the goddess of strife, a sister of Bellona. The proposed collision between these two planets was something of a family quarrel!

In creating a model of one of these planets, we have to take account of the fact that in its central region the heavier material, iron and silicates, would not be completely separated. In the Earth such separation is complete, but it has been estimated that starting with an intimate mixture of iron and silicates, complete separation would take about ten million years. Taking this into account, each of the model planets has four separate layers — a central core made of a 50 : 50 mixture of iron and silicate, a mantle with a 80 : 20 mixture of silicate and iron, a thick layer of 'ice' and an enveloping hydrogen-helium atmosphere (Figure 34.4). Ideally, in the central region the iron:silicate ratio should smoothly decrease as one moves out from the centre but in making models one must compromise between strict reality and computational practicality. We should also note that when we refer to 'ice', we mean materials such as water (H_2O), methane (CH_4), ammonia (NH_3) and carbon dioxide (CO_2) that will have been ice in the original material that produced the planet and in which any hydrogen will be deuterium rich. These materials may actually be in liquid or vapour form but we shall still call them ice. In a real planet there would not be sharp divisions between the layers. In particular, there will be ice-impregnated silicate regions and

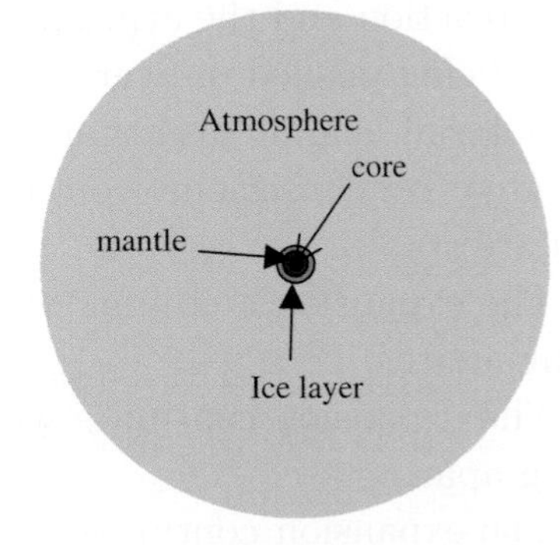

Figure 34.4. A model planet with four layers.

silicate impregnated ice regions. The ice-atmosphere boundary would certainly have been blurred by mixing of atmospheric material with ice-material vapour. The ice layer would have contained a great deal of hydrogen and this is taken with a D/H ratio of 0.01, well within the limits of observation. However, the D/H ratio in the atmosphere is taken just at the cosmic level of 2×10^{-5}.

34.6. The Collision

The progress of the collision was followed by smoothed particle hydrodynamics (SPH) previously described in Section 28.3. The way that thermonuclear reactions would influence the temperature was found by a separate calculation of a type described in Section 40.3 and the results of this calculation were built into the collision model. The stages in the collision are displayed in Figure 34.5 and are:

(a) Time = 0. The planets have just made contact.
(b) Time = 501 s. Discordia is highly distorted and material is being thrown out sideways from the collision region.
(c) Time = 1,001 s. This is similar to (b) but the highly compressed and high-temperature region is approaching the ice layer of Discordia.
(d) Time = 1,501 s. The compressed high-temperature region has reached the ice of Discordia and a thermonuclear explosion is imminent.
(e) Time = 2,003 s. A great deal of energy has been produced by the thermonuclear reaction and the atmosphere, and some of the heavier material is being pushed outwards.
(f) Time = 2,511 s. Material, mainly atmospheric, is being propelled outwards but residual condensations from the two planets have been formed and are stable.
(g) Time = 3,004 s. The expansion continues with the residual condensations moving apart.
(h) Time = 3,502 s. The expansion continues with the residual condensations moving apart.
(i) Time = 4,005 s. The expansion continues with the residual condensations moving apart.

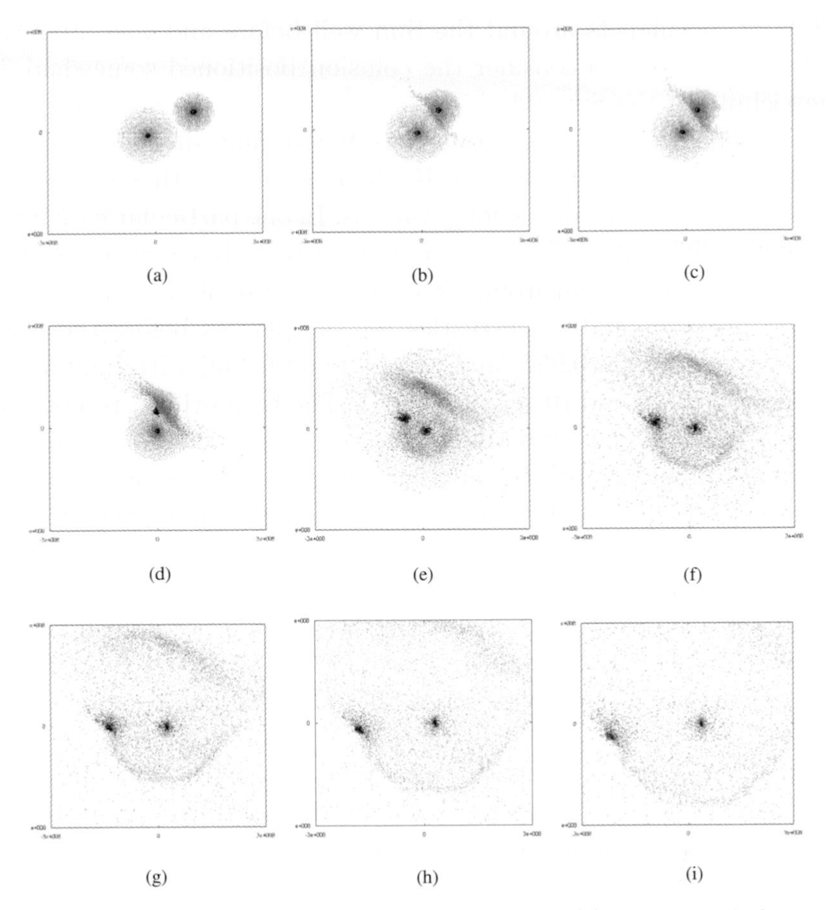

Figure 34.5. The progress of the planetary collision. (a) $t = 0$, just before contact, (b) $t = 501$ s, (c) $t = 1{,}001$ s, (d) $t = 1{,}501$ s, (e) $t = 2{,}003$ s, (f) $t = 2{,}511$ s, (g) $t = 3{,}004$ s, (h) $t = 3{,}502$ s, (i) $t = 4{,}005$ s.

An actual collision would have taken place in the presence of the Sun but the collision has been modelled as though the planets were isolated. This is quite acceptable because within the small region of the simulation, the perturbing influence of the Sun would be extremely small. To emphasize this, it should be noted that the maximum distance between the planets is just a little greater than the Earth-Moon distance and the Sun has very little influence on the orbit of the Moon around the Earth. However, if we wish to find the orbits of the planets around the Sun well before and well after the collision, we have to consider the collision positioned somewhere in the vicinity of the Sun. The relative positions of the Sun and planets do influence what the orbits were before and after the collision so that once again, all that can be done is to show that reasonable results are found for some configurations. In one particular case, with a slightly different model from that described above, with the Sun at a distance of 1.1 au from the centre of mass of the planets, the semi-major axes and eccentricities for the planets before the collision were (11.92 au, 0.957) and (3.32 au and 0.846) with final orbits (0.464 au, 0.687) and (0.881 au, 0.937). The final orbits are not precisely those of Venus and the Earth, nor would they decay to the terrestrial planetary orbits, but they are in the region of the terrestrial planets which is all that can be expected from a calculation of this type.

The final masses of the cores cannot be defined with certainty because it is not certain where to take the boundaries. In a spherical region with just over twice the radius of Venus the residue from planet Discordia contains somewhat more than the mass of Venus. In one-and-a-half times the radius of the Earth, the residue from Bellona contains 1.7 times the mass of the Earth. These are of the right order although they do not exactly reproduce the observed masses of the two larger terrestrial planets.

34.7. Summary and Comments

It has been shown that the collision of two planets, with masses in the range suggested by exoplanet observations and containing deuterium-enriched ice, starting with orbits corresponding to evolved capture-theory orbits, can give rise to stony-iron residues of the right order of mass orbiting in the region of the terrestrial planets. This suggests that Venus and the Earth may have formed in this way.

If the idea of a planetary collision explained the larger terrestrial and nothing else, then it would still be plausible since it is based on sound science and observations. However, as we shall see, it explains much more.

Chapter 35

Behold the Wandering Moon

To behold the wandering Moon,
Riding near her highest noon,
Like one that has been led astray.

John Milton (1608–1674)

35.1. Orphans of the Storm

All the major planets have considerable satellite families and there is no reason to suppose that Bellona (we take as giving the Earth) and Discordia (we take as giving Venus) were exceptional in this respect. Indeed, since Bellona was taken with twice the mass of Jupiter then it could have had satellites considerably more massive than Ganymede, the most massive of the Galileans. Given that satellites once existed for the two colliding planets, then the question arises of what became of them following the planetary collision.

There are only five possible outcomes for each of these satellites:

(i) It could have been retained by the residue of the original parent planet.
(ii) It could have changed parent and been captured by the other residue.
(iii) It could have gone into an independent orbit around the Sun.
(iv) It could have moved fast enough to have left the Solar System altogether.
(v) It could have been broken up by the collision event and its debris dispersed.

The first four outcomes can be simulated by computer calculations and outcomes (i), (iii) and (iv) are illustrated in Figure 35.1. The possible interpretations of the Moon's association with the Earth are that it was either an original satellite of Bellona that stayed with the Bellona residue (the Earth) or that it was a satellite of Discordia that attached itself to the Bellona residue. The former option is much more likely and the one we shall assume here for the purpose of discussion.

Other satellites, with other destinies, are referred to later.

35.2. A Lopsided Moon

Large astronomical bodies are usually described as spherical although they are never precisely so in practice. Because of the spins of stars and of gaseous planets they flatten along their axes and look like squashed footballs. Solid planets and satellites can take up even less symmetrical forms because of the finite strength of the material of which they are made. Thus the Moon is very slightly pear-shaped with the sharper end pointing towards the Earth. This can be understood in terms of the proposed history of the Moon. Early in its existence it would have been a hot body and somewhat more plastic in its material properties than it is now. The tidal forces due to its parent would have naturally moulded it into a slight pear-shape and at the same time ensured that the sharper end always faced the original planet. When the Moon was left in association with the residue of Bellona, the Earth, its spin period would have been different from its orbital period and all of its surface would be seen from Earth. However, eventually the combination of the shape of the Moon and the Earth's gravitational field would have locked the sharper end of the Moon's surface in the Earth-facing direction so that, presently, only the one hemisphere could be seen.

The Moon is not alone in this respect — it is a general characteristic of satellites that they orbit with one face permanently towards their parent bodies. For this reason, throughout history until recent times, we have only been able to observe that side of the Moon that

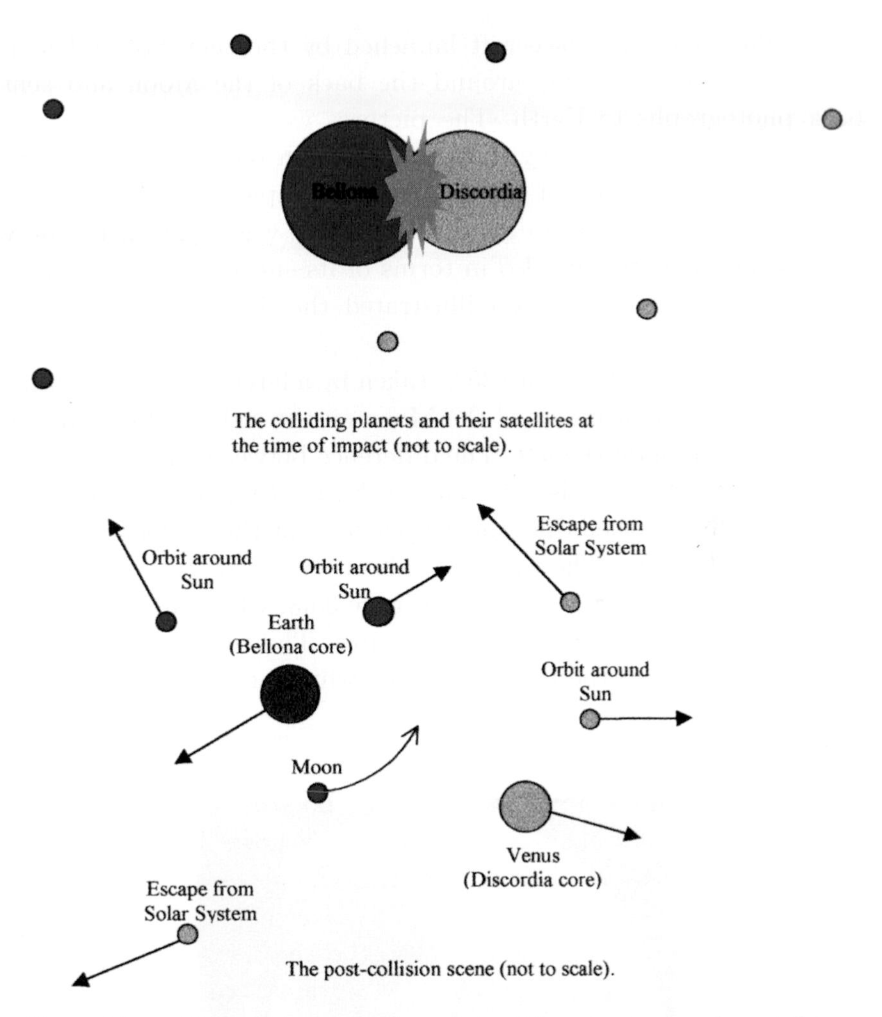

Figure 35.1. The collision scenario and the different outcomes for satellites.

faces the Earth and it was generally assumed that what we could see was a good sample of the Moon's surface as a whole. Actually, because the Moon's orbit around the Earth is in the form of an ellipse, with eccentricity 0.056 (see Figure 2.2), it appears to oscillate slightly over the course of a lunar month, a motion known as libration. This means that we can actually see 59% of the Moon's surface, although the view at the edges is very distorted.

In 1959, a Lunik spacecraft launched by the then Soviet Union made an historic journey around the back of the Moon and sent back photographs to Earth. The pictures were of poor quality by today's standards but they were good enough back then to be sensational. The rear side of the Moon was completely different from the side we could see from Earth. The Moon was lopsided not only because of its shape but also in terms of its surface features. Later, higher-quality images starkly illustrated the difference of the two hemispheres.

The photograph in Figure 35.2, taken by a later Soviet spacecraft, shows part of the near side of the Moon on the right and part of the rear of the Moon on the left. The difference between the two sides is very clear. However, it is not that the *kinds* of terrain they contain is different but rather that the *proportions* of the various kinds of terrain are different. The various kinds of terrain are well illustrated in the full-Moon picture shown in Figure 35.3, which also shows the numbered Apollo spacecraft landing sites. The two major types of surface features are the darker regions, which are *mare basins*, and

Figure 35.2. The near side (on the right) and far side of the Moon in the same image.

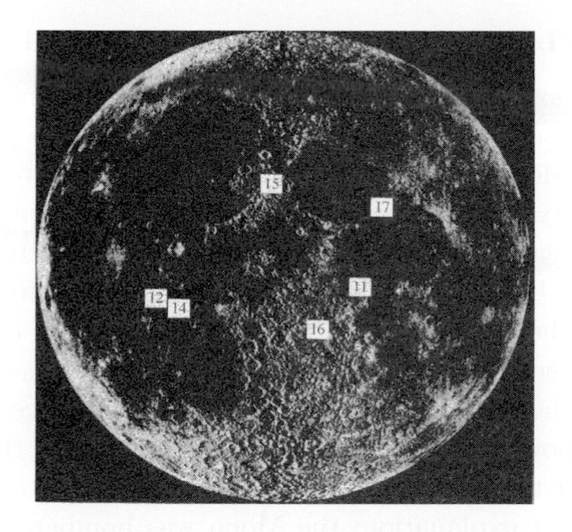

Figure 35.3. Near side of the Moon showing the Apollo landing sites.

the lighter *highland regions*. Mare basins were formed by large projectiles that struck the Moon and hollowed out what were, essentially, enormous craters. The early Moon had hot fluid material below its solid crust. The basins corresponded to weakened and thinned regions of the crust and lava was able to well up and fill the basins. Subsequently, these areas were volcanically active for several hundred million years as indicated by overlapping solidified flows of lava that can be seen in some photographs. There are relatively few smaller craters within the mare basins. This is because most such craters were produced very early in the history of the Solar System when there were many projectiles and they became obliterated by the lava flows that covered them. By contrast, the highland regions, representing areas that did not have mare basins formed within them, had no episodes of volcanism and the record of the intense bombardment, that produced a profusion of craters endlessly overlapping each other, is preserved.

The difference between the two hemispheres is simply stated. The side that faces the Earth contains many large mare basins to the extent that the terrain is dominated by these features. By contrast the far side is almost exclusively highland terrain with very few small mare basins. The question arises — why is this?

35.3. The Lopsided Moon — An Answer and a Question

An obvious explanation for the hemispherical asymmetry of the Moon would be, that for some reason or other, large projectiles fell only on the side facing the Earth. One idea put forward was that the Earth's gravity acted in such a way that it focused projectiles onto the Moon's near-surface. This theory was not supported by numerical modelling but more importantly, measurements from spacecraft showed it to be wrong. Spacecraft travelling around the Moon had radar equipment on board and by timing radar signals reflected back from the surface they were able to map the surface undulations. What this showed is that there *are* large basins on the far side of the Moon but they simply did not have lava inflows to fill them. So, there goes that explanation; the Moon was bombarded equally all over its surface.

The various Apollo missions that landed on the Moon left behind numbers of measuring devices including seismometers. These are the instruments used on Earth for recording earthquakes and, since earthquake waves travel through the Earth, the information from them enables the internal structure of the Earth to be determined. The Moon in fact is a very quiet body and such moonquakes as there are have a tiny fraction of the energy of a normal earthquake. Nevertheless, they can be recorded since the seismometers on the Moon are very sensitive. However, the seismometers can also pick up other signals — for example, due to meteorites striking the Moon's surface. In one Apollo mission, part of a Saturn rocket was deliberately targeted onto the Moon to give a good seismic signal. The outcome of these measurements is that we now know something about the Moon's internal structure. For our present purposes all we need to know is that the average thickness of the Moon's crust is about 60 kilometres, and more significantly that the crust on the near side is about 25 kilometres *thinner* than that on the far side — as illustrated in Figure 35.4.

Here then is the answer to the problem of the asymmetry of the Moon. When the Moon first formed, all its outer regions would have been molten. The less dense crust material at the surface would

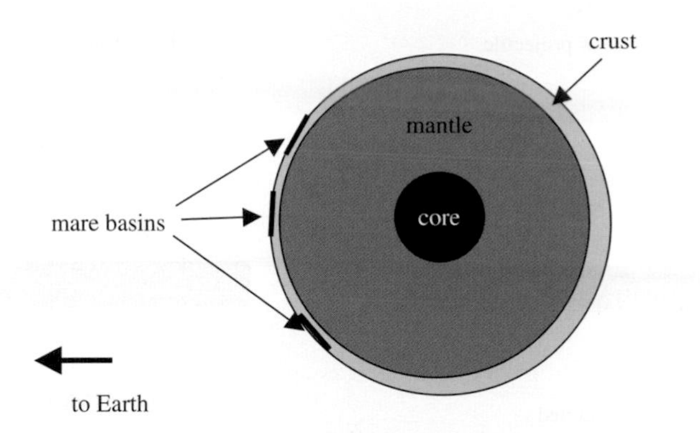

Figure 35.4. The difference of thickness of the crust on the two sides of the Moon (not to scale).

have quickly cooled and solidified and was thinner on the near side. Later, over a longish period of time, very large projectiles occasionally landed on its surface, producing basins. On the near side the crust was thin enough for lava to penetrate from below and form the mare. On the far side the lava was too far below the surface to fill the basins.

A very neat answer — but now comes the question. Why is the crust thinner on the near side? This question is all the more pertinent since it can be shown theoretically that the molten Moon, with a layer of less dense crustal material at its surface while settling down in the presence of a planet, which we take as Bellona, should have had a *thicker* crust on the near side.

35.4. Collision to the Rescue

Let us now look at the planetary collision from the Moon's point of view, taking Bellona as its original parent. The slightly pear-shaped Moon orbited Bellona with one face pointing towards its parent with a period that cannot be known but was probably between a few days to a month. Now, from a direction that did not bring it near the Moon, Discordia crashes into Bellana. Within minutes mayhem breaks loose. Debris moving at 100 km s^{-1}, or more, collides with the Moon. It is important to understand what happens when projectiles land on a solid body, such as a satellite, something which

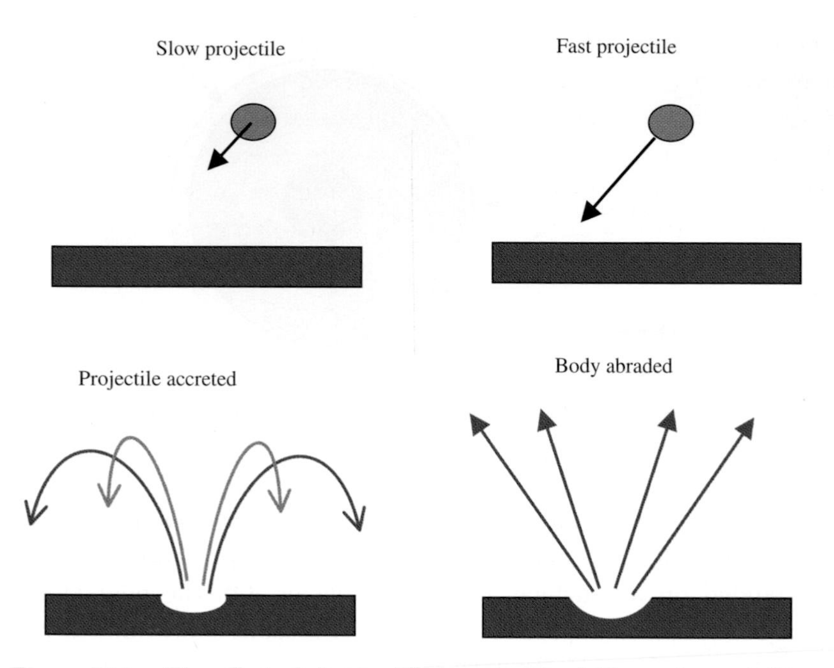

Figure 35.5. The effect of slow and fast projectiles falling onto a satellite.

is illustrated in Figure 35.5. When the projectile lands, it shares its energy with more than its own mass of surface material, some of the energy appearing as heat and the remainder as kinetic energy of surface material that flies up off the surface. If the projectile falls on the surface with a speed not too much greater than the escape speed from the body (theoretically, it cannot be less) like a meteorite falling to Earth, then none of the debris has greater than escape speed so it returns to the surface. The net result is that the *projectile is accreted by the solid body*. By contrast, if the projectile lands with a speed that is very much greater than the escape speed, then the debris, or a large part of it, can have greater than the escape speed and the *solid body is abraded by the projectile*. In the present case the escape speed from the Moon is about 2.4 km s^{-1} so the projectile lands with 40 or more times the escape speed, and hence the bombarded surface is abraded.

Now the Moon was spinning on its axis quite slowly, with the period of its spin equal to the period of its orbit, and in the period of hours during which the collision event was taking place, it would have spun very little. The rear side of the Moon was shielded from the debris and would have been essentially unaffected by the collision. The near-side material that was removed caused the thinning of its surface and calculations show that the removal of a 25 kilometres thickness of crust of the near-side is a plausible outcome.

We see that the collision really provides the answer to two questions. The first is that of how the Earth *acquired* a satellite so large in relation to its own mass. The answer given here is such that the question is actually posed in the wrong way. The Moon was *always* associated with the Earth, but the Earth itself is just the residue of a one-time planet, Bellona, most of which was dispersed. The second question relates to the asymmetric appearance of the Moon that depends on the crust being thinner on the near side. A collision provides a logical reason for this. The abrasion of the surface was sufficient to convert a hemisphere of crust that was thicker than average into one that was thinner than average.

35.5. A Brief History of the Moon

To summarize the events that have been described above, the stages in the history of the Moon are as follows:

(1) Bellona is formed in a capture event and the Moon, together with other satellites, is derived from the surrounding disk of material.
(2) The collision takes place, with the Moon being retained by the Bellona residue (the Earth).
(3) Debris abrades the hemisphere of the Moon facing the collision, converting what was originally the thickest part of the crust to become the thinnest part of the crust.
(4) The Solar System is now full of orbiting debris that occasionally collides with other bodies (mainly major planets) and is absorbed. From time to time, over the course of the next hundred million years or so, large projectiles fall on the Moon producing basins distributed all over its surface.

(5) Basins on the near side, where the crust is thinner, fill with lava from below and are also regions of volcanism over several hundred million years. Craters produced by smaller projectiles are obliterated by these flows. Craters in highland areas remain as permanent scars on the surface.
(6) As the Moon cools, the molten region moves deeper into the interior and eventually volcanism ceases. The Moon takes on its present form, unmodified except for collisions that continue to produce craters, although at a reduced rate.

Chapter 36

Fleet Mercury and Warlike Mars

Rise from the ground like a feather'd Mercury.
William Shakespeare (1564–1616), *Henry IV, Part I*

36.1. Mars as An Orphan

If the Earth and Venus cannot readily be explained as having been formed directly within the filament produced by the capture event, then this is even more so for the two small terrestrial planets, Mars and Mercury. Mars has about one-ninth, and Mercury about one-twentieth, of the mass of the Earth. To understand how these small planets fit into the general pattern of small bodies within the Solar System, it is instructive to look at the terrestrial planets and a selection of the larger satellites strung out according to their densities (Figure 36.1).

When the smaller bodies are looked at in this way, the grouping suggests that Mars can readily be associated with the rocky satellites rather than the larger terrestrial planets. We will leave consideration

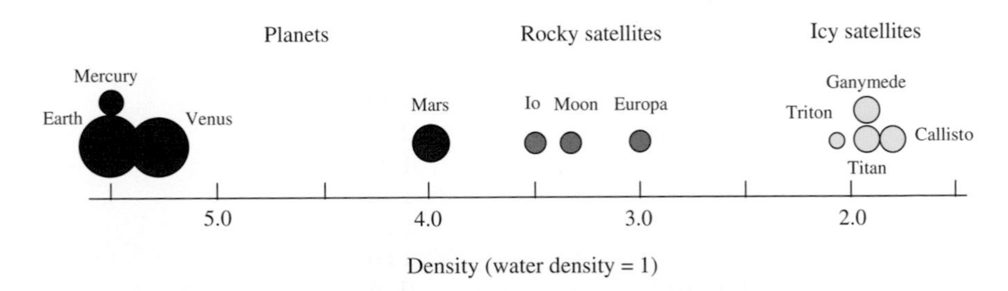

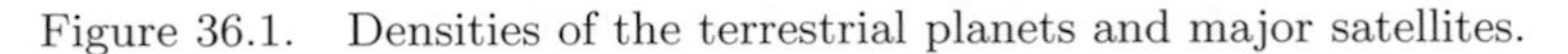
Figure 36.1. Densities of the terrestrial planets and major satellites.

of Mercury aside for the time being. Well then, how possible is it that Mars was a satellite of one of the colliding planets that was released into an orbit around the Sun that, after rounding-off and decay, ended up where it is today? It would certainly have been much more massive than the present satellites. The masses of the largest satellites and of the terrestrial planets are listed in Table 36.1 with their masses given in Moon units, and it is clear that the two small planets are considerably more massive than Ganymede, the most massive satellite. Nevertheless, they are much closer to the larger satellites in mass than to the larger terrestrial planets. However, the original Bellona was twice as massive as Jupiter, and so it could quite reasonably have had more massive satellites than any of the other planets.

Apart from the theoretical possibility that Mars could have made the transition from being a satellite to being an independent body orbiting the Sun, is there any other clue that a collision could have been involved? Yes there is. Just like the Moon, Mars shows hemispherical asymmetry as illustrated in Figure 36.2. This image of Mars shows the topography with height represented by the colours of the spectrum — red representing the highest regions and blue the lowest. The red-to-yellow highland regions in the south and

Table 36.1. The masses of the terrestrial planets and the major satellites.

Body	**Mass (Moon units)**
Earth	81.3
Venus	66.3
Mars	8.73
Mercury	4.49
Ganymede	2.04
Titan	1.93
Callisto	1.45
Io	1.21
Moon	1.00
Europa	0.66
Triton	0.30

Figure 36.2. The topography of Mars showing the northern plains and the southern highlands.

the green-to-blue northern plains show clearly the hemispherical asymmetry.

The division between the two hemispheres is marked by a scarp (a steep slope), 2 to 3 kilometres high, that runs at an angle of about 35°to the equator. The northern plains are covered with volcanic lava and contain comparatively few craters while the southern highlands, like the highland of the Moon, are heavily cratered. The deep-blue region in the south is the Hellas basin, 1,800 kilometres in diameter and 3 kilometres below the average height of Martian terrain.

All this can be explained in terms of Mars having been a satellite that was heavily bombarded on one hemisphere by the debris from colliding planets. In the case of the Moon, the crust was thinned to the extent that subsequent bombardment by huge projectiles enabled mare basins to form. However, for Mars the removal of material was so great that penetration was made closer to the subsurface molten material and extensive lava flooding of the whole region took place. This could have been due to the removal of a greater thickness of crust or to the presence of molten material closer to the surface (likely for a larger body) or even both. The southern highland region then represents the unaffected hemisphere, but this crust *was* penetrated by a massive projectile some time later to produce the Hellas basin. There is another large basin in the south, Argyre, which is 800 kilometres in diameter.

There is evidence that early in its history, Mars had a more extensive atmosphere than now and also had flowing water on its surface. It was probably an ice-covered satellite, like Jupiter's Europa, and much of this ice would have been melted and vaporized by the collision and its aftermath to produce a dense atmosphere. There would have been two main results of such an atmosphere. Firstly, due to the blanketing action of an atmosphere rich in gases such as carbon dioxide and water vapour (known as the *greenhouse effect*) the planet would have been much warmer than now so that liquid water could exist on its surface. In addition, with so much water vapour in the atmosphere, there would have been rain, and consequently, running water to form the channels that resemble riverbeds that are seen on the surface (Figure 36.3). Some of that water is now locked up in the Martian poles, another part is still present as a permafrost below the Martian surface — something that has been verified by spacecraft observations. A proposed history of the formation, and eventual

Figure 36.3. Dried up water channels on the Martian surface.

loss, of a Martian water-based climate was given by Connell and me in 1983.

An important difference between the hemispherical asymmetries of Mars and the Moon is their relationship to the spin axes or, alternatively, to the equator. The spin axis of the Moon did not change relative to surface features subsequent to the collision because it became locked to the Earth in the same configuration as it had previously been locked to the complete planet, Bellona. Consequently, the line dividing the two hemispheres runs through the lunar poles. Mars, which went into an independent orbit around the Sun was larger, much hotter and more fluid than the Moon. The solid crust was floating on a liquid mantle, and motion of the crust relative to the bulk of the planet could take place quite readily. Connell and I showed that, to satisfy a stable equilibrium condition for minimum energy, the crust would rearrange itself in such a way that mass is as far as possible from the spin axis. The present arrangement of surface features relative to the spin axis closely satisfies that requirement.

36.2. Mercury as An Orphan

Although Mars can readily be associated with rocky satellites by considering its density, that is certainly not true for Mercury. Indeed, since the density of the Earth is influenced by the compression of material due to the gravitational effect of its own mass, an effect that is negligible for Mercury, the basic *uncompressed* density of Mercury is actually greater than that of the Earth. The reason for this high density is due to a very large iron core that accounts for about 75% of the radius of the planet as a whole. The internal structure of Mercury, with that of Mars for comparison, is shown in Figure 36.4.

It will be seen that the iron core of Mercury is about the same size, perhaps a little bigger, than that of Mars. Several people have suggested that Mercury was originally a larger body that was impacted by a large projectile that stripped away a major part of its silicate outer parts so leaving it as it is now. The Capture Theory offers a scenario in which this could have occurred, although not necessarily

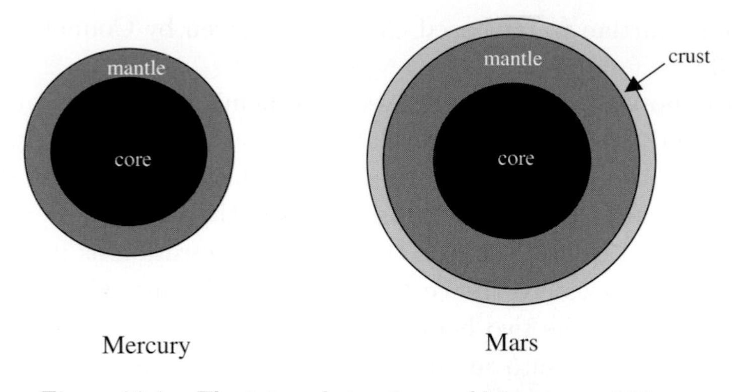

Figure 36.4. The internal structures of Mercury and Mars.

as a result of a single large projectile. If a satellite slightly larger than Mars was very close to the collision, i.e. in a near orbit to one of the colliding planets, and had some 60% of its silicate stripped away, then what was left would have a similar composition to Mercury — something I proposed in 2000. In the case of the Moon and Mars the integrity of the satellite was not totally destroyed by the abrasion process. There had to be some rearrangement of material, mainly internal, to restore equilibrium, but for the most part, the final body clearly resembled the original one. In the case of Mercury that would not have been so. Immediately after the removal of a large part of the silicate, the residue would have been an amorphous mass of material, mostly iron on the side facing the collision and mostly silicate on the other side. Gravitational forces would then have reassembled it into an iron core with an outer mantle as it is today.

The surface features of Mercury, which have been likened to those on the Moon, can be interpreted in relation to this kind of catastrophic origin. The number density of craters is less than on the Moon and there are smooth areas between the craters. A very large impact feature, the Caloris Basin, is shown in Figure 36.5. The impacting body created a series of rings that are clearly seen — produced like the ripples in a pond when a stone is thrown into it. There are also large lava plains, similar in some ways to mare features on the Moon. A feature of Mercury's surface that is not reproduced on the Moon is the presence of long and high scarps, which are wrinkles in the

Figure 36.5. Part of the surface of Mercury. The rings on the left-hand side are part of a big impact feature called the Caloris Basin.

surface due to compression as the planet cooled and shrank. This suggests that the average temperature of Mercury on formation was considerably higher than that of the Moon — compatible with the origin described here.

Chapter 37

Gods of the Sea and the Nether Regions

The moist star
Upon whose influence Neptune's empire stands...

William Shakespeare (1564–1616), *Cymbeline*

37.1. That Puny Planet Pluto

Reference has previously been made to Pluto as an odd-man-out amongst the planets — or even not a real planet at all. Everything about it points to it being aberrant in some way. Its orbital shape is the most extreme of any, with an eccentricity of 0.25. At perihelion its orbit ducks just inside that of Neptune while at aphelion, it is about 1.64 times as far from the Sun as Neptune. Not only that, its orbit, inclined at 17°to the ecliptic, has the greatest inclination of any orbit. Mercury, the nearest planet to the Sun, with an inclination of 7°is the most inclined of the other planetary orbits.

When it comes to size and mass, again it lacks the credentials to be a planet. Its diameter is less than two-thirds that of the Moon and it has barely one-sixth of the lunar mass. Only in one respect does it measure up to being a planet — it has satellites. The largest satellite, Charon, has about one-half of the diameter of Pluto while the other two are small bodies a few tens of kilometres in diameter. These satellites are in retrograde circular orbits around Pluto whose spin is also retrograde. We have already seen that not all planets have satellites so perhaps it is also possible that not all bodies with satellites are really planets. For example, the asteroid Ida is seen in

Figure 38.1 with its small satellite and by no stretch of the imagination could that body be called a planet. We are in danger of getting into a semantic maze so perhaps we should leave this line of discussion for now. Whether we choose to call Pluto a planet or not, the real goal is to explain the presence of this odd body with the characteristics it has.

The small terrestrial planets have already been explained as ex-satellites that were released into independent orbits that then rounded off and decayed to where they are now. A similar explanation could be found for Pluto. It can be postulated that it was thrown out from the collision region in an extended orbit that took it to the far reaches of the Solar System, at least beyond the orbit of Neptune. At some later stage it passed close to Neptune and was swung into something close to its present orbit (Figure 37.1). The presence of Charon could then be explained as a companion it picked up in the confused environment of the planetary collision and carried with it thereafter.

This scenario is quite plausible and fits in with the general idea of a planetary collision and its consequences. It establishes a link with Neptune, which is expected in view of the relationship of their orbits. However, it is better to look at the two bodies together to see whether they can be linked in a way that explains more than just the basic orbital characteristics of Pluto.

37.2. Neptune and its Family

Neptune itself is a perfectly normal planet, structurally similar to Uranus in many ways and with a relatively modest axial tilt of 29°.

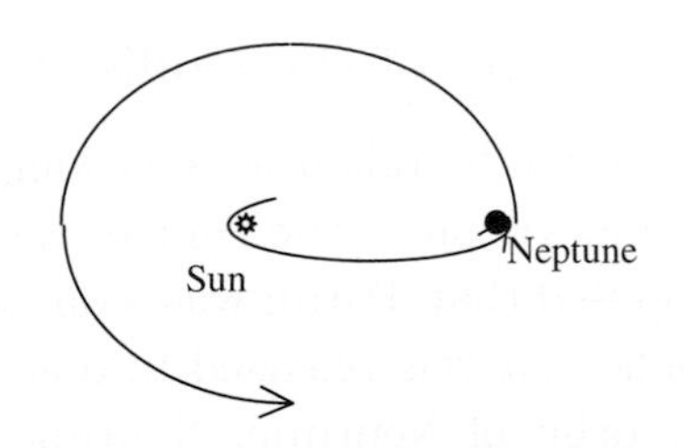

Figure 37.1. Path of Pluto deflected by Neptune. (not to scale)

However, its family of satellites shows two peculiar features that were noted by the alien Regayov in Chapter 5. To reiterate — the largest satellite, Triton, with less than one-third of the mass of the Moon, is in a *retrograde* circular orbit of radius of about 350,000 kilometres, less than the Moon's orbital radius (384,000 kilometres). A peculiarity of this retrograde orbit is that because of the way that tidal effects operate, Triton is gradually spiralling in towards Neptune. The other odd satellite, Nereid, is of modest size — a mere 340 km in diameter — and in a very extended orbit of eccentricity 0.75. The many other small satellites, that were not known before Neptune was visited by spacecraft, are all small bodies in circular orbits. One of these, Proteus, is actually slightly larger than Nereid but because it is close to the planet it was not detected from Earth observation but it was seen, along with many other small satellites, by a Voyager spacecraft.

There is yet another possible anomaly in Neptune's satellite family, not very significant in itself but perhaps yet another clue in the overall pattern that needs to be considered. Jupiter has four large satellites, including the largest of all, Ganymede. Saturn has one large satellite, Titan, which closely rivals Ganymede in mass. The largest satellite of Uranus, Titania, has less than one-twentieth of the mass of the Moon and less than one-sixth of the mass of Triton. This suggests (but not too strongly) that a satellite of Neptune with the mass of Triton does not fit the general pattern and may be an anomaly. This argument cannot be pushed too hard; the scenario that is about to be proposed only partially removes that anomaly — if it is one.

37.3. Yet Another Effect of the Collision

The starting point that will be taken for considering the relationship of Pluto to Neptune and Triton at the time of the planetary collision is as follows: It is proposed that Triton was a one-time satellite of one of the colliding planets that was released into a very extended orbit taking it beyond the orbit of Neptune. Neptune itself had a family of very ordinary satellites moving in more-or-less circular orbits. The largest of these was Pluto with a mass of one-sixth that of the Moon. In view of the general pattern of largest satellite masses with distance from the Sun, Pluto is then somewhat more massive than we might expect — but less massive than Triton. On the other hand, Uranus has had a somewhat traumatic history, as judged by its axial tilt, so maybe the problem is with Uranus, in that it should have a more massive satellite than it actually has.

We now consider the position where much of the resisting medium has evaporated and Neptune's orbit is similar to what it is now, but that is very slowly decaying. At the same time Triton is moving in its very eccentric orbit. The orbits of the two bodies will continue their differential precessions, as described in Chapter 32, and sometimes the orbits would have intersected. Now we consider an actual collision between Triton and the satellite Pluto. This was investigated by detailed computer simulation in 1999, and given suitable initial parameters, the results indicate that the system we see today could be the result of such an event.

The starting point is illustrated in Figure 37.2. Triton was in an orbit with perihelion 2.6 au and aphelion 55.6 au. Pluto was in a direct circular orbit, of radius 545,000 km around Neptune.

The collision took place with Triton on its inward journey towards the Sun. It gave Pluto a glancing blow that added energy to Pluto

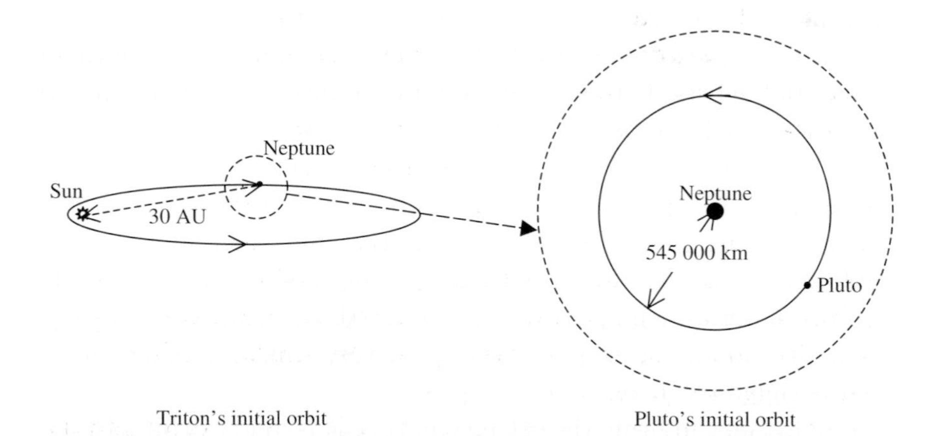

Figure 37.2. The initial orbits of Triton and Pluto before collision.

but caused it to lose energy itself. As a result Pluto was ejected from its Neptune orbit into an orbit around the Sun very similar to its present one. However, after the collision Triton was captured by Neptune into a *retrograde* orbit with semi-major axis 436,000 km and eccentricity 0.88. The tidal effect of planets on satellites with retrograde orbits is that they cause the orbits both to round-off and to decay with round-off being a fairly fast process. After the passage of time from the Triton collision event to now, the rounding-off would have certainly been complete and in fact, Triton's orbit is almost perfectly circular. The decay continues slowly.

A characteristic of orbits, in the absence of some external influence other than the Sun, is that they repeatedly go through the same parts of space. Since Neptune and Pluto were both in the same vicinity at the time of the Triton collision, it might be expected that the two orbits would continue to have parts in close proximity, although this is now not the situation for Neptune and Pluto. While their orbits overlap when seen in projection from a three-dimensional view, they are well separated. The differential precession that brought the orbits of Triton and Neptune together operated similarly to separate the orbits of Pluto and Neptune. This took place until the resisting medium was completely dispersed, at which stage the orbits became stable and separated as they now are.

One other feature of Pluto can be explained by the Triton-collision hypothesis. Triton gave Pluto a glancing blow that sheared off part of it either in the form of a single coherent large unit or in the form of substantial debris. This is the material that gave rise to Charon and the other small companions. The glancing blow was in a direction that would have spun Pluto in a retrograde direction and also led to the satellites being in a retrograde orbit — as is actually observed. It is interesting to note that a scenario that has been advanced for the formation of the Moon, by Benz, Slattery and Cameron in 1986, involves material being sheared off the Earth by collision with some body with about the mass of Mars — a very similar mechanism to what is suggested here.

Finally, we can bring the extreme nature of Nereid's orbit into the total picture. With many satellites of Neptune other than Pluto, it is likely that Triton would have passed close enough to one of them to greatly disturb its orbit. Nereid's orbit is then a manifestation of such a disturbance.

37.4. A Summary of the Triton-Collision and its Outcome

The basic scenario is that Triton was a satellite of one of the colliding planets that was released into an extended orbit around the Sun. It collided obliquely with Pluto that was in a circular direct orbit around Neptune. Pluto was expelled from its Neptune orbit into an orbit around the Sun with characteristics similar to those it now possesses. The effect of differential precession rates meant that the orbits of Neptune and Pluto drifted apart so that when the resisting medium finally disappeared, the two orbits were well-separated. Triton was captured into an elongated retrograde orbit around Neptune, which eventually rounded-off. The oblique collision gave Pluto a retrograde spin, and also removed material from Pluto that formed Charon and the other satellites orbiting Pluto in a retrograde sense.

On an incursion into the Neptune system, which need not have been the one where the collision with Pluto took place, Triton passed close enough to Nereid, a satellite previously in a more-or-less circular orbit, and perturbed it into its present highly eccentric orbit.

Chapter 38

Bits and Pieces

Gather up the fragments that remain, that nothing be lost.

The Gospel according to St John

38.1. The Gap and its Denizens

When the alien Regayov (Chapter 5) explored the Solar System he noted that there appeared to be a gap between Mars and Jupiter. He then found, on closer approach, that this gap was occupied by many thousands of the small bodies that we know as asteroids. The asteroids mostly move between Mars and Jupiter but some of them have orbits that take them within the orbit of Mars, or even within that of the Earth, and some are well outside the orbit of Jupiter. They are all in direct orbits around the Sun, mostly with eccentricity less than about 0.4 and orbital inclinations less than about 30°, although a few have more extreme orbital characteristics.

Asteroids come in a large range of sizes, from a diameter of just under a thousand kilometres downward. It is impossible to state what the smallest size is, but they have a continuous distribution of sizes down to objects so small we cannot reasonably call them asteroids any more. The smaller bodies, that are several metres or less in dimension, occur in great profusion and many of them fall onto the Earth. These bodies are *meteorites* and they are highly treasured by scientists for whom they are rich sources of information. Virtually all meteorites have been produced as the debris from the occasional collisions of asteroids although a few are thought to be fragments knocked off the surface of Mars. An examination of the composition and properties of meteorites is tantamount to similar studies of the asteroids from which they were derived. In fact, by comparing the way that meteorites reflect light in laboratory experiments with the way that asteroids reflect sunlight, it is possible to associate asteroids with meteorites of various types.

Some important asteroids, together with their orbital characteristics are listed in Table 38.1.

Ceres is similar in size to some of the moderate-size satellites of the major planets. Eros, Apollo and Icarus are "Earth-crossing asteroids" that move within the Earth's orbit in part of their paths. It is possible that an Earth-crossing asteroid could collide with the Earth — but it is not a possibility that should cause us loss of sleep. Icarus has an orbit of extreme eccentricity that takes it very close to the Sun. Finally, Chiron has nearly all its orbit in the region between Saturn and Uranus.

Objects larger than about 400 kilometres in diameter take up a spherical shape, or nearly so. For such large objects, the force of gravity tries to pull all the material inwards as far as possible. For smaller objects, the strength of the material will resist a change of shape but larger objects will be forced towards a spherical form. Ceres is close to spherical but the majority of the smaller asteroids

Table 38.1. Characteristics of some important asteroids.

Asteroid	**Year of discovery**	**Semi-major axis (au)**	**Diameter (km)**	**Eccentricity**	**Inclination (°)**
Ceres	1801	2.75	933	0.079	11
Pallas	1802	2.77	608	0.237	35
Juno	1804	2.67	250	0.257	13
Vesta	1807	2.58	538	0.089	7
Hygeia	1849	3.15	450	0.100	4
Undina	1867	3.20	250	0.072	10
Eros	1898	1.46	25	0.223	11
Hildago	1920	5.81	15	0.657	43
Apollo	1932	1.47	?	0.566	6
Icarus	1949	1.08	$\sim$2	0.827	23
Chiron	1977	13.50	?	0.378	7

Figure 38.1. The asteroid Ida with its satellite, Dactyl.

are of irregular shape. This is illustrated in the NASA spacecraft picture of Ida, an asteroid with dimensions of about $56 \times 24 \times 21$ km, which has the peculiar feature that it has a small satellite (Dactyl), seen as a tiny dot on the right-hand-side of Figure 38.1.

38.2. Some Ideas on the Origin of Asteroids

Long before images of asteroids were available, it was known that they were irregular in shape. Asteroids tumble in space with periods in the order of two to four hours and as they do, their brightness changes according to how much area they present to the Sun's illumination and to the observer. The irregular shapes of asteroids suggested to many planetary scientists that they were the result of some event that broke up a larger body.

An early idea about the origin of asteroids was that they represented the debris from a planet that once existed in the region of the asteroid belt, but that it somehow exploded. There are various difficulties with this idea. One is that the total mass of all known asteroids is very small, a small fraction of the mass of the Moon, so it would be necessary to explain where the remaining part of the planet had gone. This could be explained by saying that, over a long period of time, the debris had been swept up by collisions with planets, particularly Jupiter. Another, perhaps more serious, difficulty is that there is no known source of energy that could cause a planet to spontaneously explode and completely disintegrate.

An alternative explanation that has been advanced for asteroids is that planets were produced by the gradual accumulation of bodies similar to asteroids so that asteroids are just the unused material from this process. However, there are also difficulties with this idea. Firstly, the physical composition of some meteorites shows that at some point in time, their material had been molten. Secondly, meteorites consist of two main types — irons and stones — reflecting the type of material that comprises them, with very few containing substantial proportions of both iron and stone. Both these pieces of evidence suggest that asteroids were derived from larger bodies that were molten and within which a substantial gravitational field existed that would separate material into layers of different densities. If material is brought together to form a large body, then gravitational energy is released in the form of heat. Theory shows that if a spherical body of mass M is assembled by bringing material together from a large distance away, then the heat energy released is

$$H = C\rho M^{5/3}, \tag{38.1}$$

where ρ is the density of the material and C is a constant. Thus, the amount of heat generated per unit mass of material is

$$H/M = C\rho M^{2/3} \tag{38.2}$$

which increases with increasing M. For this reason the accumulation of a sufficiently massive body can give melting of its material. It can be shown that for stony material, the critical radius is about 1,200 kilometres, much greater than that of Ceres but smaller than that of the Moon. We have already noted that the Moon was molten when it was young, so giving the volcanism that was discussed in Chapter 35. Within a body of a size that would give melting, material would segregate through gravity to give an iron core and a stony mantle, thus giving the separation of material seen in meteorites and inferred in asteroids. It seems unlikely that bodies the size of asteroids could melt and produce a segregation of iron and silicate and this problem has given rise to the idea of *parent bodies*, the collisions and disruptions of which produced asteroids. According to SNT workers, these parent bodies are produced by accumulations of planetesimals (Section 20.3). It is suggested that parent bodies, smaller than the critical mass for melting through gravitational energy release, could have melted due to the presence of a radioactive isotope of aluminium when they first formed. This isotope, aluminium-26, has a short half-life of about 720,000 years, which means that it produced all its heating effect within a few million years and that it has now completely disappeared. Actually, if aluminium-26 was around in the early Solar System in the quantities suggested, then even small asteroids, a few kilometres in diameter, could have melted.

38.3. The Planetary Collision Again!

What would the colliding planets have been like? We have postulated that Bellona may have originally been twice as massive as Jupiter and Discordia somewhat more massive than Saturn. It is expected that the cores of the colliding planets would have been similar to those of the present major planets in the Solar System. We recall the description of the gas giants given in Section 34.5. At the very centre there is an iron core surrounded by a silicate mantle, which put together has a few times the mass of the Earth. Moving outwards towards the outer parts of the mantle, there would have been an increasing component of volatile materials of various kinds — predominantly water, carbon dioxide, ammonia and methane — materials that would form ice when cooled. If such material cooled, the mixture of silicates and ice would resemble permafrost on Earth, or as the amount of silicate decreased, a frozen marsh. There would then have been a region of volatiles without silicates and finally a thick hydrogen + helium atmosphere with admixtures of small amounts of other gases. The proposed structure is shown in Figure 38.2 where the volatiles are labelled 'ice'.

The temperature generated in the region of the impact was a few million degrees and very quickly went much higher, for reasons

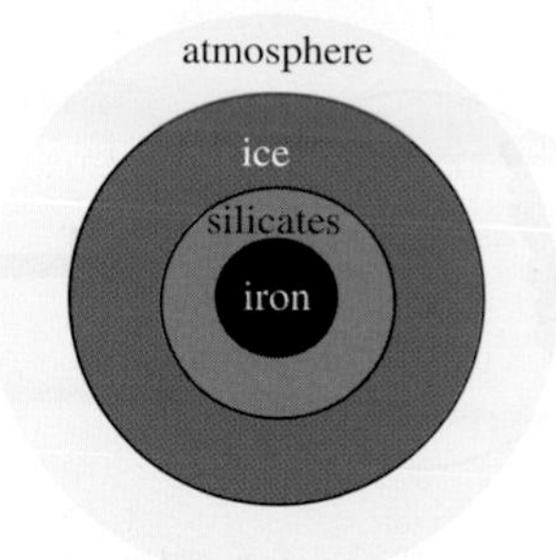

Figure 38.2. A model colliding planet (regions not to scale). The silicate-ice shells are not sharply delineated, as shown, but merge into one another in a gradual way.

that were explained in Section 34.4. However, in regions of the planets remote from the collision interface, the temperature would have increased much less.

The break-up of the planets involved several processes. The atmospheres eventually completely dispersed, partly by being heated and partly because of the loss of the material that gravitationally bound them to the planets. Stony material that was vaporized then cooled and formed small molten droplets that were incorporated within other debris that had not been heated to such high temperatures. Large quantities of molten material remained as coherent masses and then cooled and solidified to form various types of asteroid. In regions remote from the collision, large chunks of solid material, varying from almost pure silicate through to silicates highly impregnated with ice, were ejected from the planets by a process called *spallation* (Figure 38.3).This is a process that depends on energy being transmitted through a body by shock waves. One way of showing this on Earth is to take a frangible object like a brick, to hold it firmly in a vice and to strike it at one end with a hammer. Shock waves travelling through the brick cause flakes at the other end to break off and to be ejected.

The spallation process propelled the outermost material of the planetary cores outwards with greater energy than that acquired by deeper material and hence to greater distances. Some material ejected in this way may even have completely escaped from the Solar System.

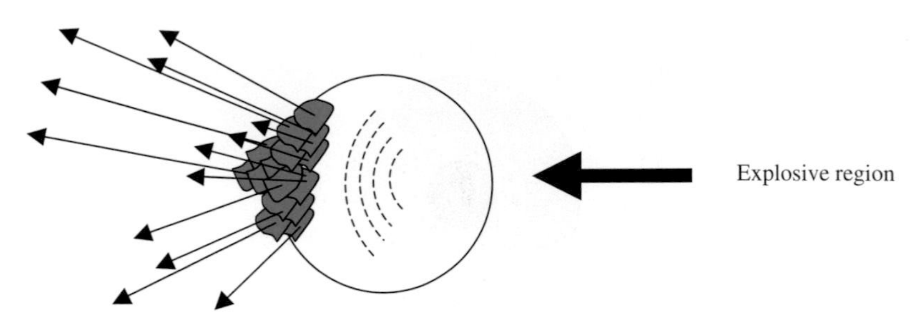

Figure 38.3. Material spalled off the far sides of the colliding planet cores by shock waves moving through them.

Given this set of mechanisms, we examine the evidence from the present objects in the Solar System to see whether or not the pattern of what is observed is compatible with the collision hypothesis.

38.4. How do we Interpret Meteorites?

Although the iron cores of the planets were shielded by many other layers of material, some molten iron globs would have been ejected, especially from the side of the core closest to the collision region. These would first have solidified and then eventually cooled to form iron asteroids or iron regions within mixed-material asteroids. The iron in planets has a significant admixture (a few percent) of nickel and when the iron is solid, but hot, the nickel moves about within the iron to form different nickel-iron compounds. Examination of sections of iron meteorites show interesting features called Widmanstätten figures, shown in Figure 38.4, that come about because iron and nickel separates out into two different alloys, *taenite*, which is rich in nickel, and *kamacite*, which contains less nickel.

From the sizes of the features of the Widmanstätten pattern it is possible to estimate the rate at which the material cooled, and estimates of cooling rates are usually in the range 1–10 K per million years. Since small objects cool quickly and large objects cool slowly this rate of cooling gives an idea of the sizes of the cooling bodies and it can be deduced that the cooling took place in asteroid-size bodies.

Figure 38.4. Widmanstätten figures.

Stone meteorites are of three main kinds. The first kind is called *chondrites*, so called because the meteorites contain chondrules, which are small glassy spheres. The chondrule shown in Figure 38.5 is about 1 mm across.

These glassy spheres are frozen droplets of once-molten silicates, the sort of material that would be produced when a silicate vapour cools and condenses. In addition, the various minerals found in the droplets indicate that they froze very quickly. When the chondrule was molten, or even just very hot, the groups of atoms within it could move around freely and they would reassemble themselves into an assembly of very stable minerals, whatever the initial mineral composition of the chondrule was (Section 18.1). However, the minerals

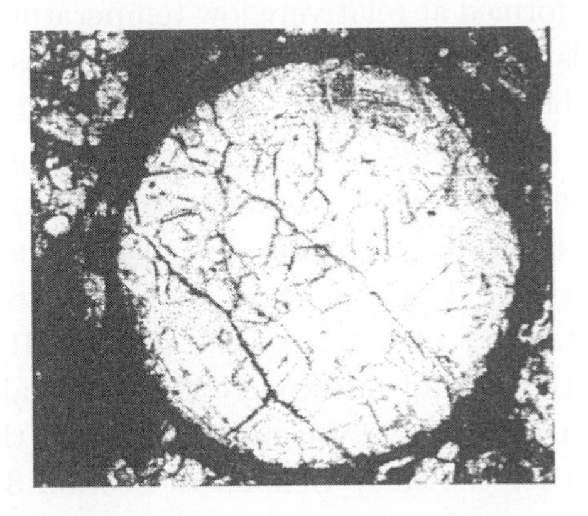

Figure 38.5. A typical chondrule.

found in chondrules do *not* correspond to these very stable minerals, from which it is deduced that they cooled very quickly, hence stopping movement of the groups of atoms before they could reorganize themselves into the favoured composition. Such rapid cooling, or *quenching*, is just the expectation from a planetary collision. Although a planetary collision may be thought of as a massive event from a human perspective, from an astronomical point of view, it is quite small scale and the ejection of material and its subsequent cooling would have taken place on a timescale of minutes.

The second kind of stone meteorite is the *achondrites*, so-called because they have no chondrules (the prefix *a* in Greek means *without*). In many ways, these resemble the surface rocks of terrestrial planets and some of them are very similar to terrestrial materials. They can be related to the silicate material, coming from the colliding planets, which was not greatly heated and survived in much its original form.

The third kind of stone meteorite is *carbonaceous chondrites*, some of which, but not all, contain chondrules. They are notable for the amount of volatile materials they contain, including carbon compounds such as benzene. They also contain water — up to 22% in some types — although the water is bound into minerals and is not present in liquid form. Many carbonaceous chondrites contain small white inclusions consisting of minerals that condensed at very high temperatures. This is odd because the main part of these meteorites must have formed at relatively low temperatures, otherwise the volatile materials would have been driven off. It can be interpreted as there having been condensation of very hot material to form the white inclusions that were later incorporated into cooler material that made up the bulk of the carbonaceous chondrites.

Finally, there are some meteorites designated as stony-irons. They form two main types. The first type is the *pallasites* where there are silicate crystals set in a framework of iron. It has very much the appearance that would be expected of material derived from a region of a planet where there was some mixing of the two materials as we postulated for the early Bellona in Section 34.5. The second type is *mesosiderites*. Here, there is a chaotic mixture of iron and stone fragments with the iron sometimes in globules and sometimes as veins within the stony regions. Some of the stony minerals are of a form that could not exist at the high pressures in the deep interior of a planet. These are readily interpreted as admixtures of stony and iron material coming together in the aftermath of the collision. Some of the minerals would be from surface regions and so never subjected to high pressures thus explaining this characteristic of mesosiderites.

38.5. How do we Interpret Asteroids?

In terms of their general shapes and sizes it is clearly quite plausible that asteroids are the debris of a collision, consisting of objects of different composition, sizes and sources from within the original planets. Otherwise, what we know about asteroids as objects is confined to what we can infer by associating them with different kinds of meteorite, as judged by similarities in the way they reflect light. About 75% of all asteroids fall into the *C-type* category meaning that they resemble carbonaceous chondrites. Another 17% are *S-type* and are presumed to be of stony-iron character and there are a few *M-type* that are more highly reflective and thought to be of iron-nickel composition.

The general pattern is that on average, the C-type asteroids are further from the Sun than the other types although there is quite a large spread of locations of each type. This is consistent with expectation from a collision where material with a higher volatile content would be further out in the planet and hence expelled with greater speed to greater distances from the Sun.

38.6. A Summary

Asteroids and, particularly, meteorites are a rich source of information about conditions that occurred in the early Solar System. Their features have been interpreted in terms of many different scenarios but what is shown here is that the collision hypothesis is consistent with what is observed and seems to be able to account for the main features of these bodies.

Chapter 39

Comets — The Harbingers of Doom!

When beggars die there are no comets seen,
The heavens themselves blaze forth the death of princes.

William Shakespeare (1564–1616), *Julius Caesar*

39.1. Early Superstition

The quotation heading this chapter comes from Shakespeare's Julius Caesar. Caesar's wife, Calphurnia, had noted the appearance of a comet and she was appealing to Caesar not to attend the Senate that day. This play was set in Roman times but the superstition that the excerpt reveals was still prevalent in the Elizabethan era. Comets can be awesome apparitions, spanning the sky and dwarfing all other astronomical phenomena. It was no wonder that they were taken as the portent of great events — usually disasters. The appearance of Halley's comet in 1066 was, no doubt, a disastrous omen for King Harold — although William of Normandy probably took a rather different view!

39.2. What is a Comet?

It was Newton's friend, Edmund Halley, who first recognized that comets were bodies in orbit around the Sun and should therefore repeat their appearances in a periodic way. He suggested that the comet seen in 1682, which now bears his name, was one that had previously been seen in 1607, 1531 and 1456 and that it would reappear in 1758. He did not live to see his prediction come true. However, since the comet had a period of about 76 years, Halley deduced from Newton's work that the comet had a semi-major axis of 18 au. This result, coupled with the knowledge that the comet went close to the Sun, was the first indication that there were some bodies that travelled around the Sun in orbits of high eccentricity.

The essential component of a comet is a relatively small nucleus (diameter of a few kilometres) consisting of silicates heavily impregnated with various ices — principally water ice and frozen carbon dioxide and ammonia. At distances from the Sun beyond the orbit of Jupiter, comets are very difficult to see since they are dark inert objects. As they approach the Sun, they undergo a dramatic transformation. Heat from the Sun vaporizes material within them that escapes from the surface and forms a ball, called the *coma*, around the nucleus. The action of sunlight on these vapours is to break them up chemically and the components of the break-up fluoresce. Fluorescence is the phenomenon where a substance absorbs radiation over a large range of wavelengths but then re-emits the energy as radiation at very specific wavelengths that are a characteristic of the material. Thus, by analysing the light from the coma it can be ascertained that it contains a large number of compounds mostly consisting of combinations of hydrogen, carbon, nitrogen and oxygen.

What is seen at the head of comet West, in Figure 39.1, is the coma. Its radius is about one hundred thousand kilometres, which makes it larger than Jupiter. Enveloping the coma is an even larger cloud of hydrogen, with up to ten times its radius. This does not emit visible light so it cannot be seen by the naked eye, but it can be detected with scientific instruments.

The most spectacular part of a comet is its tail (Figure 39.2). The Sun is constantly emitting large numbers of charged particles, mostly protons and electrons, which travel outwards at high speed. This so-called *solar wind* bombards the material of the coma and makes some of it stream out in a direction away from the Sun. Due to the action of solar radiation the vapour in the coma is broken down into

Figure 39.1. Comet West (1976).

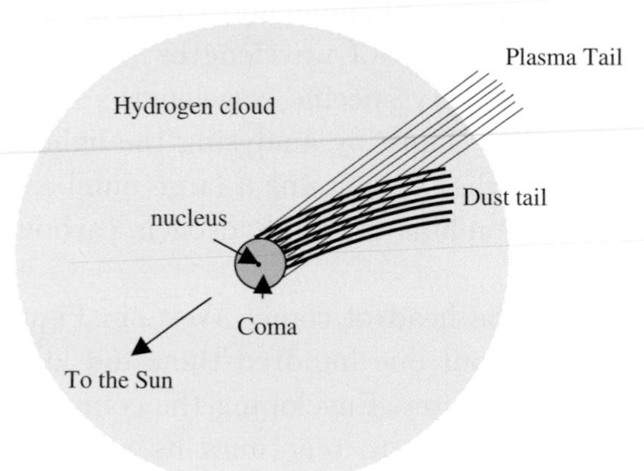

Figure 39.2. A schematic representation of the structure of a comet.

a plasma, an intimate mixture of negatively charged electrons and positively charged combinations of atoms. The coma also contains small dust particles that are prised off the comet by the jet effect of the escaping vapour. These two components of the coma sometimes form two distinctive tails, something that can be seen in the image

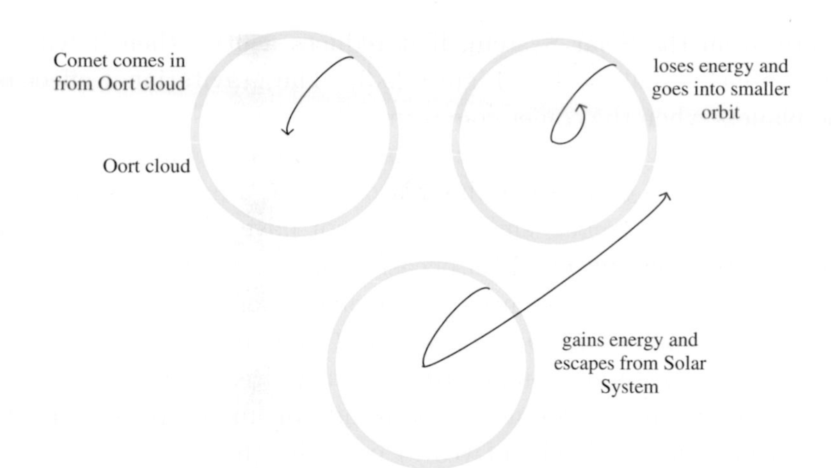

Figure 39.3. The effect of adding or subtracting a small amount of energy from an Oort cloud comet.

of comet West. The upper thinner tail is the *plasma tail* and the thicker and stubbier lower one is the *dust tail*.

One very famous American astronomer, Fred Whipple, once described a comet as a 'dirty snowball' and that general description is as good as any.

39.3. The Different Kinds of Comet Orbit

Halley's comet, with an orbital period of about 76 years, has its orbit contained in the region occupied by the planets. Any comet with a period of 200 years or less will have all or most of its orbit within the planetary region and these are called *short-period comets*. There are about 100 comets in this category. However, even these comets can be subdivided; those with orbital periods of more than about 20 years can be in either direct or retrograde orbits so that, for example, Halley's orbit is retrograde. Comets with periods less than 20 years are all in direct orbits.

Just under once per year a comet appears that comes from a distance of tens of thousands of au so that it is in an extremely eccentric orbit. An orbit like this is very sensitive to a small change in its energy. If the energy increases a little then it will be able to escape from the Solar System, if it reduces a little then it will go into a much smaller orbit (Figure 39.3). The gravitational effect of the planets when the comet comes into the central Solar System will change its energy by an amount that is almost bound to give one or other of these effects. For this reason, these comets from afar are described as *new comets* since they cannot have entered the central Solar System on any previous occasion on their existing orbit. An implication from the existence of new comets was given by Jan Oort, a Dutch astronomer, in 1948. He deduced that surrounding the Solar System there is a vast cloud of 10^{11} (one hundred thousand million) comets at distances of tens of thousands of au (corresponding to orbital periods of millions of years) in orbits that keep them away from the planetary region of the Solar System. This system is now known as the *Oort cloud*. Once in a while a passing star, or some other perturbing influence, will cause some of these comets to change their orbits into ones that will pass close enough to the Sun for them to be observed. This is the way that new comets are produced.

39.4. Some Problems with Comets

Every time a comet goes close to the Sun and becomes active it loses some of its inventory of volatile materials. After 1,000 orbits or so the volatiles are exhausted, and once that happens, the comet becomes an inert dark body and essentially disappears from view. Since the average lifetime of a short period comet is about 10,000 years and there are about 100 of them this means that, to maintain the population of visible comets requires a new one to be created every 100 years or so. It was first thought that the short-period comets were produced when new comets happened to be perturbed by one of the major planets in such a way that its new orbit stayed within the planetary region. Calculations showed that this would be a very inefficient process and could do nothing like maintain short-period comets in their current number.

In 1951, Gerard Kuiper (1905–1973), a Dutch-born, but later American, astronomer, suggested that outside the orbit of Neptune there should be a population of small bodies in direct orbits around the Sun. His reason for this idea was that he could not accept that the Solar System would simply come to an abrupt end at the distance of the outer planets. Later, it was realised that if such a population existed, it would solve the conundrum of the source of short-period comets. Bodies in this region would be perturbed, particularly by Neptune, and once in a while one would stray close to Neptune and be thrown into the inner Solar System. In 1992, the first such body outside Neptune, at a distance of 40 au from the Sun, was discovered and since then many hundreds more have been detected. This region of bodies is now known as the *Kuiper belt* and it has been estimated that it could contain up to 70,000 bodies with diameters of 100 km or more and a much larger number of smaller bodies that could form potential comet nuclei.

Kuiper

Another problem that has exercised planetary scientists is the continued existence of the Oort cloud. It has been mentioned that the occasional passage of a star could perturb the Oort cloud to give new comets. Simple calculations show that during the lifetime of the Solar System, about four and a half billion (thousand million) years, some stars and other massive astronomical objects must have come so close that they would have virtually swept away the Oort cloud. So why is it still there?

The resolution of this problem is bound up with the existence of the Kuiper belt. The suggestion has been made that there is a vast number of potential comets in an inner cloud, concentrated at distances from hundreds to one or two thousand astronomical units from the Sun but stretching out much further. The Kuiper belt is then interpreted as the extreme inner fringe of this belt. When a near passage of a star or other perturbing body takes place, it does indeed remove many members of the Oort cloud. However, at the same time, perturbation of bodies further in takes place and those that are driven outwards reinforce the depleted Oort cloud to maintain its existence.

The theory hangs together quite well. Given the establishment of an inner belt of comet-like bodies the Oort cloud, Kuiper belt and short-period comets fit into an inter-linked pattern.

39.5. Yes, You Guessed It — The Planetary Collision Again

The planetary collision gave the outcome that material closest to the solid surface of the planet was the most volatile and was also that which was thrown out the fastest. Material ejected from near-surface regions would have ended up most remote from the Sun and would have extended outwards over a wide range of distances. Some of it may even have been ejected from the Solar System. Now, while it seems that this provides an explanation for the inner comet belt, it only does so if some other factors are taken into account. Consider a portion of this debris thrown out to a large distance from the collision region at, say, 1 au from the Sun. Orbits are closed. i.e. they repeat their journeys through the same points in space, so eventually the piece of debris comes back into the inner Solar System. If this happens, then its chance of survival over the lifetime of the Solar System is very small and it will not become a member of the inner belt. But now, take into consideration that the collision took place when protoplanets were on very extended orbits. For such orbits, the protoplanets spent most of the time at great distances from the Sun. If the debris passed near one of the planets, say at 500 au from the Sun, then it could have been thrown into an orbit that kept it well

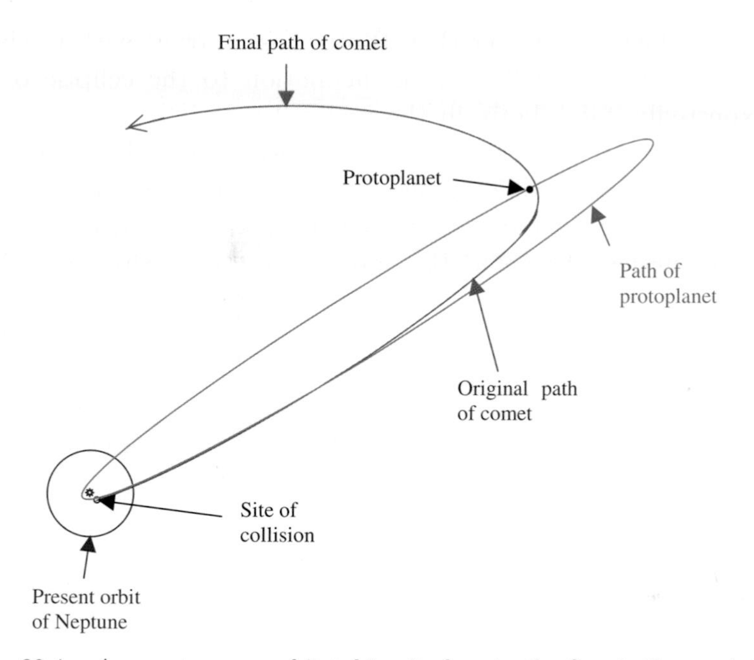

Figure 39.4. A comet on an orbit taking it close to the Sun is thrown into an orbit well outside the present planetary region by interaction with a planet in its early extended orbit.

outside the present 40 au limit for planets, as is illustrated in Figure 39.4. It could also remain outside the main region of the resisting medium so that its orbit would hardly have decayed. It then needed to survive for a million years, by which time the protoplanets would all have decayed inwards, and thereafter it would have been safe.

Most of the debris would not have the good fortune to survive in this way but even a tiny survival rate would explain what we have today. The total mass of all the present comet material is probably considerably less than one Earth mass but the total mass of released ice-impregnated silicates could have been several Earth masses. Thus a survival rate of a few percent is all that is needed to account for what now exists.

The process we have just described for creating the inner reservoir of comets also serves well for explaining the existence of Eris, the so-called tenth planet discovered in 2005. This object, some 2,800 km in diameter, making it larger than Pluto, has an orbit with perihelion 76 au, eccentricity 0.85 and an inclination to the ecliptic of 44°. Unexpectedly, for a body in the region beyond the Kuiper belt it is not mainly ice and both water and methane are absent. If it had been a satellite of a colliding planet, thrown out to a great distance, which interacted with an early planet still on an extended orbit, then both its physical characteristics and its orbit are entirely explicable.

Chapter 40

Making Atoms With a Biggish Bang

I can trace my ancestry back to a protoplasmal primordial atomic globule.

W. S. Gilbert (1836–1911)

40.1. Let's Find Out More About Isotopes

Sometimes in a newspaper or magazine article, the word 'isotope' turns up — often in the context of something that is made in a nuclear reactor and has some medical use. Well then — what is an isotope? In Figure 25.1, we showed a representation of a carbon atom, the one labelled carbon-12 in Figure 40.1. The nucleus contains six protons (red), each with a positive charge and six uncharged neutrons (black). There are also six negatively charged electrons (blue) in the region around the nucleus to balance out the proton charges and to make the atom as a whole electrically neutral.

Electrons have a very tiny mass that can be disregarded in respect to the total mass of the atom. Protons and neutrons have masses that

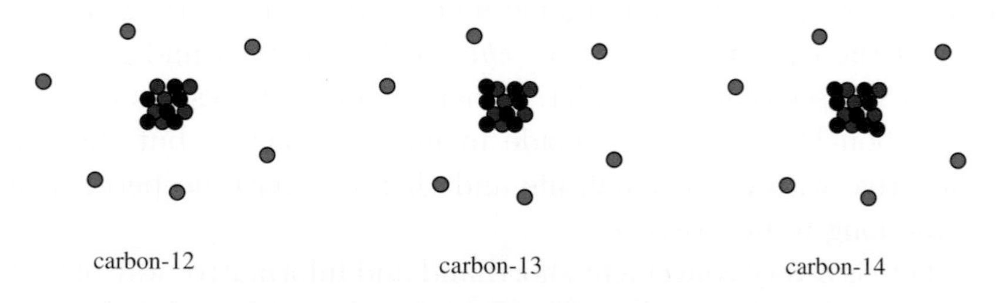

Figure 40.1. The structures of carbon-12, carbon-13 and carbon-14.

are nearly equal and 1,840 times greater than that of the electron. The '12' in the designation 'carbon-12' means that the atom has 12 units of mass. We do not have to specify the number of protons (equal to the number of electrons) because it is six protons in the nucleus that makes the atom carbon and not some other atom. Most living material is made of carbon, and that includes us. If we could examine the individual carbon atoms in our bodies, *most* of them would be carbon-12. However, just over 1% of them would be different — we would find that the nucleus contains seven neutrons, not six (Figure 40.1). Is it still carbon? Yes it is — it is a perfectly normal carbon atom (the six protons ensure that) and it will behave just like any other carbon atom in a chemical experiment. It is an *isotope* of carbon called carbon-13 because the atom has 13 units of mass. Just over 1% of all the carbon atoms on Earth, from whatever source, are carbon-13.

Actually a very diligent search amongst all these carbon atoms would find another isotope, carbon-14, that has eight neutrons in its nucleus. It is, however, a somewhat uncomfortable atom and would rather be something else. Carbon-14 is *radioactive* and once in a while, apparently at random, a carbon-14 atom undergoes a transition that turns it into a nitrogen atom. If there is a whole collection of carbon-14 atoms, then in 5,900 years one half of them will have transformed into nitrogen; we say that the *half-life* of carbon-14 is 5,900 years. Despite its short half-life compared with the age of the Solar System, carbon-14 occurs naturally since it is continuously produced by the action of cosmic rays in the upper layers of the atmosphere. All living matter contains carbon-14 and archaeologists use it to date specimens of dead organic matter. It must be stressed again that all the carbon isotopes are *chemically* identical and it is only in their masses that they differ. Other carbon isotopes are possible, e.g. carbon-11, and can be made in nuclear reactors but they are radioactive with a short half-life and therefore have no permanent, or even long term, existence.

There is a very convenient shorthand and informative way of writing carbon-12, which is $^{12}_{6}\mathrm{C}$. The C is the *chemical symbol* that represents carbon. The numbers, 12 and 6, give the mass of the atom (number of protons + neutrons) and the number of protons respectively. Actually, the six is redundant and scientists, who know how many protons there are in a carbon atom, will simply write ^{12}C. However, we shall continue to use the redundant form. These, then, are the isotopes of carbon we have considered so far —

$$^{12}_{6}C \; ^{13}_{6}C \; ^{14}_{6}C$$

Here is a summary of what we have learnt above. For many atoms there is more than one stable isotope. There may also be naturally occurring radioactive isotopes, which change into some other kind of atom, with a particular half-life. There is one other fact that we should know. When we examine carbon on Earth, from any source, we find the ratio $^{12}_{6}C : ^{13}_{6}C = 89.9 : 1$. If we find an object with a significantly different ratio of carbon isotopes, then we may reasonably infer that it is not of terrestrial origin. A similar statement can be made about the isotopic ratios of other atoms. For many atoms, there is a terrestrial isotopic composition and any specimen giving a different composition is probably of non-terrestrial origin.

40.2. Isotopes in Meteorites

We certainly know that meteorites have not come from Earth but are mostly bits knocked off from asteroids. Nevertheless, while the model that we have for producing planets would not necessarily make them all the same chemically, we might expect them all to be the same in terms of their isotopic composition. If we found a meteorite with $^{12}_{6}C : ^{13}_{6}C$ very different from $89.9 : 1$ then we might wonder what had happened to its material, or indeed, whether it was really from the same original source as terrestrial material. It takes a lot to change isotopic composition. Small changes can occur, due to heating effects or other physical conditions, but they would be both small and explicable.

Actually, meteorites show very marked differences of isotopic composition from terrestrial material for many different kinds of atom. Here we shall just deal with a selection of such differences, which are usually designated as *isotopic anomalies*.

40.2.1. *The carbon anomaly*

In some chondritic meteorites (Chapter 38) there exist grains of the mineral silicon carbide, chemically represented as SiC. We already know that C is the chemical symbol for carbon and Si is the symbol for silicon. The chemical representation shows us that one atom of silicon combines with one atom of carbon to form a silicon carbide unit. When the ratio ${}^{12}_{6}\mathrm{C} : {}^{13}_{6}\mathrm{C}$ is found for the carbon in samples of SiC from different meteorites, they are found to vary greatly and many are much less than 89.9 — down to 20, or even less. There is more of the heavier isotope ${}^{13}_{6}\mathrm{C}$ than is found on Earth and this isotopic anomaly is referred to as *heavy carbon*.

An explanation that has been given for heavy carbon is that it is the result of an influx of material into the Solar System from distant carbon stars. These are stars that are known, from analysis of the light they emit, to be rich in carbon — and the carbon they contain is heavy carbon. It has been suggested that the various isotopic compositions of SiC grains can be explained if they originated in six or more carbon stars. The grains then drifted across interstellar space and were incorporated into meteorites.

40.2.2. *The oxygen anomaly*

The normal oxygen (chemical symbol O) that we breathe consists of three isotopes represented as ${}^{16}_{8}\mathrm{O}$, ${}^{17}_{8}\mathrm{O}$ and ${}^{18}_{8}\mathrm{O}$. You should now be in a position to know how many protons, neutrons and electrons those isotopes contain. The standard terrestrial ratios for those isotopes are ${}^{16}_{8}\mathrm{O} : {}^{17}_{8}\mathrm{O} : {}^{18}_{8}\mathrm{O} = 0.953 : 0.007 : 0.040$ and this mixture is known as SMOW (Standard Mean Ocean Water). Given that various thermal processes can change those ratios a little, but in a predictable way, oxygen specimens from the Earth and Moon seem quite compatible.

Samples of oxygen from some ordinary chondrites and carbonaceous chondrites give oxygen isotopic ratios that cannot be explained by simple physical processing of normal terrestrial oxygen. However, they *can* be interpreted as being admixtures of SMOW with different amounts of pure ${}^{16}_{8}\mathrm{O}$, but the problem is to find some source of pure, or nearly pure, ${}^{16}_{8}\mathrm{O}$. One explanation that has been offered is that it is produced by the action of alpha particles (the nuclei of helium atoms, represented by $^{4}_{2}$He) on $^{12}_{6}$C in distant stars. This pure $^{16}_{8}$O is then incorporated into dust grains that travel to the Solar System and become incorporated in meteorites. Subsequently, normal solar-system oxygen infiltrated the grains displacing most, but not all, of the $^{16}_{8}$O thereby giving a final content that is rich in that oxygen isotope.

40.2.3. *The magnesium anomaly*

There are three stable isotopes of magnesium (chemical symbol Mg) in the ratios

$$^{24}_{12}\mathrm{Mg} : {}^{25}_{12}\mathrm{Mg} : {}^{26}_{12}\mathrm{Mg} = 0.79 : 0.10 : 0.11.$$

The white high-temperature inclusions in carbonaceous chondrites (Section 38.4) were found to contain excess $^{26}_{12}$Mg in proportion to the amount of aluminium in the sample. There is only one stable isotope of aluminium, $^{27}_{13}$Al, but, as mentioned in relation to the melting of asteroids (Section 38.2), there is the radioactive isotope of aluminium, $^{26}_{13}$Al, with a half-life of 720,000 years. When it decays it leaves behind $^{26}_{12}$Mg.

The explanation given for the magnesium anomaly is that when the meteorite formed, the aluminium within it contained a small fractional component of $^{26}_{13}$Al — of the order of one part in 10,000 or 100,000. This then decayed leaving an excess of $^{26}_{12}$Mg in the meteorite. Different mineral grains in a meteorite would have different chemical compositions but the aluminium in them will all be derived from the same source and hence have the same ratio of $^{26}_{13}\mathrm{Al} : {}^{27}_{13}\mathrm{Al}$. The excess of $^{26}_{12}$Mg in a particular mineral grain within the meteorite would then be proportional to the amount of aluminium (hence also of $^{26}_{13}$Al) that it contained.

In Chapter 26, we mentioned that the event that triggered the formation of a cool dense cloud was a supernova and this would have produced a wide range of isotopic products, including $^{26}_{13}$Al. However, it takes about 200 million years for the cool dense cloud to be produced by which time all the $^{26}_{13}$Al would have disappeared. To explain the evidence for $^{26}_{13}$Al being present in the early solar system it has been suggested that just before the Solar System formed, there was a second supernova somewhere in its vicinity.

40.2.4. *The neon anomaly*

Neon (chemical symbol Ne) forms a small proportion of the air we breathe and it is best known as the material that gives rise to the orange glow of 'neon signs'. It is chemically inert, which means that it does not normally produce chemical compounds, therefore when it occurs in meteorites it is in the form of individual neon atoms. These are trapped in atomic-size cavities but they can be released by heating the meteorite. In fact, if neon, or some other gas, is found in meteorites it is certain that these meteorites were not substantially heated after the neon was incorporated — otherwise the neon would have escaped.

Normal terrestrial neon has the following composition

$$^{20}_{10}\mathrm{Ne} : {}^{21}_{10}\mathrm{Ne} : {}^{22}_{10}\mathrm{Ne} = 0.905 : 0.003 : 0.092.$$

Specimens of neon from different solar-system sources are very variable in isotopic composition and it has been deduced that there may be three separate sources, and that the neon we normally measure in meteorites is formed from different admixtures of these sources.

There is another kind of neon in meteorites that cannot be explained as a mixture of three basic components. Some meteorites contain pure, or almost pure, neon-22, which is a 9% component of terrestrial neon. This anomalous neon is called neon-E. There is no conceivable natural mechanism for completely separating neon-22 from a mixture of isotopes and so some other explanation is required. The most plausible source of this neon is by the decay of a radioactive isotope of sodium (chemical symbol Na), $^{22}_{11}$Na. The only stable isotope of sodium is $^{23}_{11}$Na. One scenario that has been suggested is that $^{22}_{11}$Na was produced in a supernova and then was incorporated, together with stable sodium, in minerals. The $^{22}_{11}$Na decayed and the resultant $^{22}_{10}$Ne was trapped within the mineral grain. The only difficulty with this scenario is that $^{22}_{11}$Na has a very short half-life — 2.6

years. This means that the radioactive sodium has to be produced in a supernova and then incorporated into a *cool* solid body within a period of 10-20 years. If the body were not cool at the time of its incorporation, then the neon would not have been retained.

40.2.5. *Hydrogen Isotopes*

In Section 34.2, we discussed the various hydrogen isotopes and discovered why it is that deuterium is heavily concentrated in the ices of cool dark clouds, protostars and, presumably, early protoplanets. Spacecraft have been able to collect micron-sized particles, called interplanetary dust particles (IDPs), from the region above the stratosphere, which prove to be rich in deuterium with D/H up to 0.01. A possible explanation of IDPs is that they represent micron-size mineral grains originally trapped in the otherwise pure ice (mostly water, methane and ammonia) that formed the outermost solid layer of one or both of the colliding planets. Hydrated minerals, for example, serpentine that is highly hydrated and is an important component of some carbonaceous chondrites, would exhibit the D/H ratio of the ices within which they existed.

40.3. For The Last Time — A Planetary Collision

We now take the major planets that collided as having had regions containing hydrogen-containing molecules enriched in deuterium with D/H at the level suggested by IDPs and lower than that allowed by the observations described in Section 34.3. When the planets collided, the temperature due to the collision process itself was about 2–4 million K and we consider the possibility that nuclear reactions could have taken place. While a temperature of, say, 3 million K may seem very high, in terms of setting off nuclear reactions it is actually quite low. In fact, there is only one kind of nuclear reaction that can be triggered at such a low temperature and that is one involving two deuterium atoms — and we have a good bit of that around. Once a nuclear reaction process is triggered in any local region there could be a snowball effect — as long as the concentration of deuterium is sufficient to raise the local temperature to about one hundred million

K, at which value other nuclear reactions can take place. Thereafter, nuclear reactions generate high temperature and high temperature generates nuclear reactions. In one calculation, Holden and I took the starting materials as those that would be expected in a mix of icy silicates with hydrogen-rich molecular material. A triggering temperature of three million K was applied and the calculation included 568 different kinds of nuclear reaction involving different isotopes. The reactions went into an explosive mode and the temperature generated was 500 million K.

The outcome of this calculation was that we were able to explain all the isotopic anomalies described in this chapter, plus others we have not dealt with here.

Carbon

Heavy carbon was produced covering the complete range of observed values.

Oxygen

In the heart of the explosion, the reactions that took place virtually destroyed all the ${}^{17}_{8}O$ and ${}^{18}_{8}O$ but left the ${}^{16}_{8}O$ intact. This provided the source of pure ${}^{16}_{8}O$ that mixed with other normal oxygen to give the anomalous oxygen.

Magnesium

A large quantity of ${}^{26}_{13}Al$ was produced, sufficient to explain the magnesium anomaly.

Neon

A sufficient quantity of ${}^{22}_{11}Na$ was produced to explain the production of neon-E. We noted the idea that ${}^{22}_{11}Na$ was produced in a supernova and was then incorporated in cold material on a timescale of 10-20 years. This is somewhat more plausible for production in a planetary collision. The scale of a planetary collision is small by astronomical standards and it can be shown that expansion, the formation of grains and cooling would all take place within hours or days.

40.4. Deuterium in the Colliding Planets and Other Bodies

The production of isotopic anomalies depended on the reactions that took place in a nuclear explosion. The nuclear explosion depended on reaching a high-enough temperature to initiate it and that temperature depended on the amount of deuterium available. The minimum D/H ratio required within an icy environment to set off other nuclear reactions is about 0.005, i.e. about one half of that indicated by IDPs.

We have proposed that comets and asteroids (hence meteorites) were derived from those parts of the core and mantle regions of the colliding planets that were remote from the exploding region. These bodies show evidence of enhanced D/H ratios (Section 34.2). Some material would have been mostly ice containing small quantities of silicate. When the ice evaporated, there would have been a residue of silicate fragments, including fine dust that would be in the form of IDPs. The deuterium bound into hydrated minerals in IDPs would be more resistant to exchange with external hydrogen than that would be in volatile molecules and so, as expected, IDPs show a higher D/H ratio than do comets.

Chapter 41

Is the Capture Theory Valid?

If God were to hold out enclosed in His right hand all Truth, and in His left hand just the active search for truth, although with the condition that I should ever err within, and should say to me: Choose! I should humbly take his left hand and say: Father! Give me this one; absolute truth belongs to Thee alone.

G.E. Lessing (1729–1781)

The answer to the question that forms the subject of this chapter was given in Chapter 1 — no theory can ever be designated as true. This is not just being evasive, like not answering the question 'Will you beat your wife again?' with a simple "yes" or "no". In this case, the question is meaningless because it is impossible in principle to know what the answer is. If the question is changed to 'Is the theory plausible?' then there is the possibility of an answer without evasion.

None of the mechanisms that have been described, from the initial cooling of the interstellar medium to the tidal and impact interactions that gave various features of the planets, is outside the boundaries of what science will comfortably allow. In addition, the probabilities of the various events have all been shown to be acceptably high. One has to be a little cautious with probability arguments. As human beings, we are all individually highly improbable. The probability that the particular combination of parental genes was formed that produced a particular individual is incredibly low. However, given the statistics of reproduction, someone has to be born. For the model described here for the initial planetary orbits, the probability that Bellona and

Discordia collided is quite small — let us say 0.01. Remember that in considering this probability, there was a period of about a million years in which the two planets could have collided. With the very eccentric and extensive orbits of the six initial major planets there were fifteen possible pairs of planets that could have collided. If the probability for each pair was 0.01, then the probability of *some* collision event was $0.01 \times 15 = 0.15$ — so not at all unlikely. The numbers are only illustrative. An actual calculation for a slightly different model showed that some major event was more likely than not, but for that calculation the definition of a major event included a close interaction in which one or other of the planets was thrown out of the Solar System.

So it is claimed that the Capture Theory is *plausible* — no more, no less. That is something that has to be left to the judgement of the reader and scientists who can understand the theoretical basis of what has been described. But even if it were agreed that, as related so far, the theory gave *perfect* explanations for the origin and evolution of the Solar System that would still not exclude the possibility that some new evidence could discredit it completely. Nor can the possibility be excluded that some other theory, based on a completely different scenario, could be advanced that would be equally plausible. Having said that, experience with this subject, stretching in the scientific era of over 200 years, suggests that it is not easy to find good explanations for the wide variety of features of the Solar System. Any theory that appears to do so should not be lightly dismissed.

The main test to be applied to decide whether or not a theory should be taken seriously is to consider its consistency and coherence — or lack of it. Does the story flow with a continuity of events like the narration of a long novel — or is it a disjointed series of short stories without obvious links? Does it involve a minimum number of assumptions and events to explain the diverse features of the system? This last point is really a reiteration of Occam's razor, as mentioned in Chapter 1. A theory that gives a series of *ad hoc* explanations for almost every feature of the Solar System, lacking any common root, would not inspire confidence. For example, in Chapter 40, we saw that many anomalous isotopic features have been assumed by meteoriticists to have come from various sources outside the Solar System. The planetary collision model, and the associated nuclear explosion, derives all the anomalies in one event, supporting evidence which exists in other features of the Solar System as well.

The account given here divides into two parts. The first describes the origin of the Solar System from the initial interstellar medium through to the formation of a set of major planets with satellites. The second part is concerned with the evolution of the initial system. It starts with planets, accompanied by satellites, in highly elliptical but evolving orbits and ends with the Solar System in much the state we see it today. The two parts are intimately linked in that for the evolutionary processes to be able to take place, it is necessary to have the starting point provided by the first part. There must be planets in highly elliptical orbits undergoing precession, otherwise a collision cannot occur. It is from that one event, the collision, that so much else follows. Of course, if some other model can be found for the formation of planets that also leads to major planets in similar highly eccentric orbits, then the second evolutionary part of this model could still apply.

The strength of the model is that sequences of events are causally related. From cooling of the interstellar medium we have cool, dense collapsing clouds. Turbulence in the clouds gives the collisions of turbulent elements that lead to stars. The stars, falling inwards in a collapsing cloud, give the dense embedded stage where interactions between condensed stars and either protostars or dense compressed regions take place and planets are formed. The collapsing planets develop disks that provide material for satellite formation. The same kind of sequencing can be made for the second part that also has another feature — a single event, the planetary collision that leads to explanations of features as different as the origin and characteristics of Triton and isotopic anomalies in meteorites.

The evolutionary structure of the model is illustrated below in the form of a simple flow diagram with colour coding to show related features. It should again be stressed that all the really critical features of the model, and some of the less critical features, have been subjected to detailed modelling. It should also be said that not everything that comes out of the Capture Theory approach has been included — for example, ideas about lunar magnetism — but such excluded details are very peripheral to the main description.

Naturally, the theory may not be correct and eventually there may be a better one.

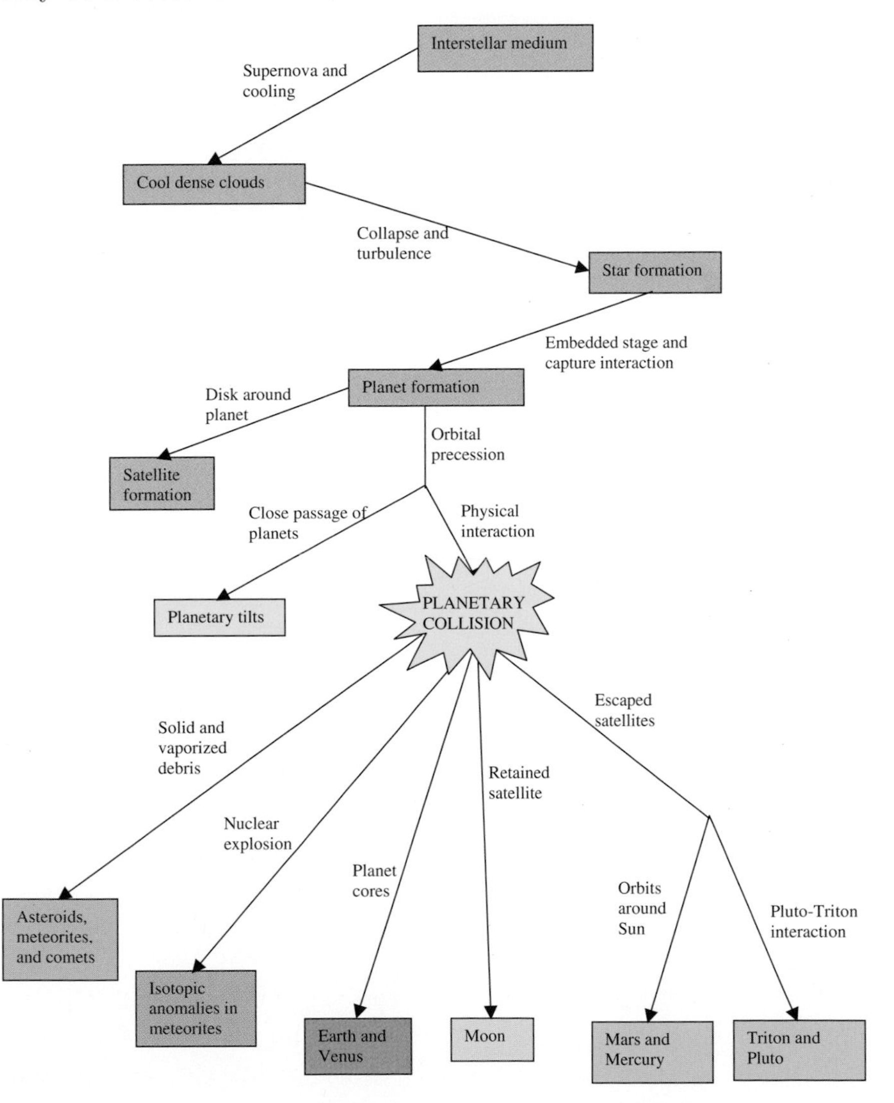

Epilogue

An Autumn Evening

Greg settled himself into the armchair with a contented sigh and gazed into the log fire. Lydia had excelled herself that evening; in particular the venison, his favourite treat, had been cooked to perfection. She was now settling the kids into bed and she would soon bring in the coffee to round off what had been the perfect meal. Then they would both settle down to read — not talking, but still enjoying each other's company.

He glanced at the coffee table at his elbow on which was the journal that had just arrived and which contained the papers presented at the spring conference on '*The Origin of the Solar System*'. He had not presented a paper himself but Henry, his research student, had given a talk on '*The Detection of Interstellar Dust within the Solar System*' that had been very well received. It was an excellent meeting and, despite the many problems that remained, he had a sense that real progress was made. He recalled that the latest work on the top-down formation of planets had provoked a lively debate but it was only by considering and discussing all possible ideas that a realistic model would eventually emerge. After all, everyone agreed on the basic model — that the Solar System, and planetary systems in general, had grown out of a nebula — and it was just the details of the formation of the various types of body that had to be worked out. Actually, he recalled, perhaps not everyone agreed on the nebula model. There was one participant who had raised a question about some alternative model involving a tidal interaction between stars — or something like that — but the Chairman had commented, in his rather acerbic way, that they were there to discuss serious science and could not spend valuable time talking about offbeat ideas. The Chairman also reminded the audience that tidal theories had been investigated and found wanting almost one hundred years before and there was little point in trying to revive them now. At the time, it had occurred to Greg that the Chairman was being a little inconsistent here — after all, nebula ideas had been revived after being abandoned for nearly a century.

Lydia arrived with the coffee. Greg picked up his journal and began to read.

Bibliography

Chapter 7

Laplace, P. S. de (1796) *Exposition du Système du Monde* (Imprimerie Cercle-Social, Paris).

Roche, E. (1873) Essai sur la constitution et l'origine du système solaire, *Mem. Acad. Montpellier* **8**, pp. 235–324.

Chapter 8

Chamberlin, T. C. and Moulton, F. R. (1900) Certain recent attempts to test the nebula hypothesis, *Science* **12**, pp. 201–208.

Moulton, F. R. (1907) On the probability of a near approach of two suns and the orbits of material ejected from them under the stimulus of their mutual tidal disturbances, *Carn. Inst. Yb.* **5**, pp. 168–169.

Nölke, F. (1908) *Das Problem der Entwicklung unseres Planetensystems. Aufstellung eine neuen Theorie nach vorgehende Kritik der Theorien von Kant, Laplace, Poincaré, Moulton, Arrhenius, u.a.* (Springer, Berlin).

Chapter 9

Jeans, J. H. (1917) The motions of tidally distorted masses with special reference to theories of cosmogony, *Memoirs of the Royal Astronomical Society* **652**, pp. 1–48.

Jeans, J. H. (1938) The origin of the planets, *Science and Culture* **4**, pp. 73–75.

Jeans, J. H. (1942) Origin of the Solar System, *Nature* **149**, p. 695.

Jeffreys, H. (1929) Collision and the origin of rotation in the Solar System, *M. N. Roy. Astron. Soc. Supplement* **89**, pp. 636–641.

Russell, H. N. (1935) *The Solar System and its Origin* (Macmillan, New York).

Spitzer, L. (1939) The dissipation of planetary filaments, *Astrophys. J.* **90**, pp. 675–688.

Chapter 10

Aust, C and Woolfson, M. M. (1973) On the accretion mechanism for the formation of a protoplanetary disk, *M.N. Roy. Astron. Soc.* **161**, pp. 7–13.

Bondi, H and Hoyle, F. (1944) On the accretion of matter by stars, *M. N. Roy. Astron. Soc.* **104**, pp. 273–282.

Lyttleton, R. (1960) Dynamical calculations relating to the origin of the Solar System, *M.N. Roy. Astron. Soc.* **121**, pp. 551–569.

Schmidt, O. J. (1944) A meteoric theory of the origin of the Earth and planets, *Comptes Rendus (Doklady) Academie des Sciences de l'URSS* **45**, pp. 229–233.

Chapter 11

Jeffreys, H. (1952) The origin of the Solar System, *Proc. Roy. Soc.* **A214**, pp. 281–291.

Weizsacker, C. H. von (1944) Űber die Entstehung des Planetensystems, *Z. fur Astrophysik* **22**, pp. 319–355.

Chapter 12

McCrea, W. H. (1960) The origin of the Solar System, *Proc. Roy. Soc.* **A256**, pp. 245–266.

McCrea, W. H. (1988) Formation of the Solar System: Brief review and revised protoplanet theory, in *The Physics of the Planets: Their Origin, Evolution and Structure*, ed. S. K. Runcorn (Wiley, Chichester).

Chapter 14

Balli, J., Testi, L., Sargent, A. and Carlstrom, J. (1998) Lifetimes of externally illuminated young stellar objects embedded in the Orion Nebula, *Astron. J.* **116**, pp. 854–859.

Haisch, K. E., Lada, E. A. and Lada, C. J. (2001) Disk frequencies and lifetimes in young clusters, *Astrophys. J.* **553**, pp. L153–L156.

Chapter 15

Butler, R. P. and Marcy, G. W. (1996) A planet orbiting 47 Ursae Majoris, *Astrophys. J.* **464**, pp. L153–L156.

Cameron, A. C. (2001) Extrasolar planets, *Physics World.*

Chapter 16

Holland, W. S., Greaves, T. S., Zuckerman, B., *et al.* (1998) Submillimetre images of dusty debris around nearby stars, *Nature* **392**, pp. 788–790.

Chapter 17

Brush, S. G. (1996) *Fruitful Encounters: The Origin of the Solar System and of the Moon from Chamberlin to Apollo* (Cambridge University Press, Cambridge).

Chapter 18

Armitage, P. J. and Clarke, C. J. (1996) Magnetic braking of T Tauri stars, *M. N. Roy. Astron Soc.* **280**, pp. 458–468.

Hoyle, F. (1960) On the origin of the Solar Nebula, *Q. J. R. Astr. Soc.* **1**, pp. 28–55.

Lynden-Bell, D. and Pringle, J. E. (1974) The evolution of viscous disks and the origin of nebula variables, *M. N. Roy. Astron. Soc.* **168**, pp. 603–637.

Chapter 19

Cameron, A. G. W. (1978) The primitive solar accretion disk and the formation of planets, *The Origin of the Solar System*. Ed. S. F. Dermott (Wiley, Chichester).

Chapter 20

Blum, J., Wurm, G., Kempf, S., *et al.* (2000) On growth and form of planetary seedlings, *Phys. Rev. Lett.* **85**, pp. 2426–2429.

Goldreich, P. and Ward, W. R. (1973) The formation of planetesimals. *Astrophys. J.* **183**, pp. 1051–1061.

Konacki, M. (2005) An extrasolar giant planet in a close triple-star system, *Nature* **436**, pp. 230–233.

Safronov, V. S. (1972) *Evolution of the Protoplanetary Cloud and Formation of the Earth and Planets* (Israel Program for Scientific Translations, Jerusalem).

Weidenschilling, S. J., Donn, B. and Meakin, P. (1989) *The Formation and Evolution of Planetary Systems*, eds. H. A. Weaver and L. Danley (Cambridge University Press, Cambridge), pp. 131–150.

Chapter 21

Wetherill, G. W. and Stewart, G. R. (1989) Accumulation of a swarm of small planetesimals, *Icarus* **77**, pp. 330–357.

Wetherill, G. W. and Stewart, G. R. (1993) Formation of planetary embryos: Effects of fragmentation, low relative velocity, and independent variation of eccentricity and inclination, *Icarus* **106**, pp. 190–209.

Chapter 22

Cole, G. H. A. and Woolfson, M. M. (2002) *Planetary Science: The Science of Planets around Stars* (Institute of Physics, Bristol).

Ward, W. R. (1997) in ASP Conf. Ser. eds. Rettig, T. and Hahn, J. M., *On the formation and migration of protoplanets*, Proc. 1st Int. Origins Conf. Astron. Soc. Pac. (San Francisco).

Chapter 23

Beckwith, S. V. W. and Sargent, A. (1996) Circumstellar disks and the search for neighbouring planetary systems, *Nature* **383**, pp. 139–144.

Boss, A. P. (2000) Possible rapid gas giant formation in the solar nebula and other protoplanetary disks, *Astrophys. J.* **536**, pp. L101–L104.

Chapter 25

Golanski, Y. and Woolfson, M. M. (2001) A smooth particle hydrodynamics simulation of the collapse of the interstellar medium, *M. N. Roy. Astron. Soc.* **320**, pp. 1–11.

Goldsmith, D. W., Habig, H and Field, G. B. (1969) Thermal properties of interstellar gas heated by cosmic rays, *Astrophys. J.* **158**, pp. 173–184.

Hayashi, C. (1966) Evolution of Protostars, *Ann. Revs. Astron. Astrophys.* **4**, pp. 171–192.

Chapter 26

Bonnell, I. A., Clarke, C. J., Bate, M. R. and Pringle, J. E. (2001) Accretion in stellar clusters and the initial mass function, *M. N. Roy. Astron. Soc.* **324**, pp. 573–579.

Krumholz, M. R., McKee, C. F. and Klein, R. I. (2005) The formation of stars by gravitational collapse rather than competitive accretion, *Nature* **438**, pp. 332–334.

Woolfson, M. M. (1979) Star formation in a galactic cluster, *Phil. Trans. R. Soc. Lond.* **A291**, pp. 219–252.

Chapter 27

Kroupa, P. (1995) Star cluster evolution, dynamical age estimation and the kinematic signature of star formation, *M. N. Roy. Astron. Soc.* **277**, pp. 1522–1540.

Lada, C. J. and Lada, E. A. (2003) Embedded clusters in molecular clouds, *Ann. Rev. Astron. Astrophys.* **41**, pp. 57–115.

Chapter 28

Oxley, S. and Woolfson, M. M. (2003) Smoothed particle hydrodynamics with radiation transfer, *M. N. Roy. Astron. Soc.* **343**, pp. 900–912.

Oxley, S. and Woolfson, M. M. (2004) The formation of planetary systems, *M. N. Roy. Astron. Soc.* **348**, pp. 1135–1149.

Woolfson, M. M. (1964) A capture theory of the origin of the solar system, *Proc. R. Soc.* **A282**, pp. 485–507.

Chapter 29

Dantona, F. and Mazzitelli, I. (1998) http//www.mporzio.astro.it/~dantona/prems.html.

Melita, M. D. and Woolfson, M. M. (1996) Planetary commensurabilities driven by accretion and dynamical friction, *M. N. Roy. Astron. Soc.* **280**, pp. 854–862.

Woolfson, M. M. (2003) The evolution of eccentric protoplanetary orbits, *M. N. Roy. Astron. Soc.* **340**, pp. 43–51.

Chapter 30

Woolfson, M. M. (2004) The stability of evolving planetary orbits in an embedded cluster, *M. N. Roy. Astron. Soc.* **348**, pp. 1150–1156.

Chapter 31

Schofield, N. and Woolfson, M. M. (1982) The early evolution of Jupiter in the absence of solar tidal forces, *M. N. Roy. Astron. Soc.* **198**, pp. 947–959.

Woolfson, M. M. (2004) The formation of regular satellites, *M. N. Roy. Astron. Soc.* **384**, pp. 419–426.

Chapter 32

Cole, G. H. A. and Woolfson, M. M. (2002) *Planetary Science: The Science of Planets around Stars* (Institute of Physics, Bristol).

Dormand, J. R. and Woolfson, M. M. (1977) Interactions in the early solar system, *M. N. Roy. Astron. Soc.* **180**, pp. 243–279.

Woolfson, M. M. (2000) *The Origin and Evolution of the Solar System* (Institute of Physics, Bristol).

Chapter 33

Hutchison, R. (1983) *A Search for our Beginning* (Oxford University Press, Oxford).

Chapter 34

Dormand, J. R. and Woolfson, M. M. (1977) Interactions in the early solar system, *M. N. Roy. Astron. Soc.* **180**, pp. 243–279.

Parise, B., Caux, E., Castets, A., *et al.* (2005) HDO abundance in the envelope of the solar-type protostar IRAS 16293–2422, *Astron & Astrophys* **431**, pp. 547–554.

Parise, B., Ceccarelli, C., Tielens, *et al.* (2002) Detection of doubly-deuterated methanol in the solar-type protostar IRAS 16293–2422, *Astron & Astrophys* **393**, pp. L49–L53.

Roberts, H., Herbst, E. and Millar, T. J. (2003) Enhanced deuterium fractionation in dense interstellar cores resulting from multiply deuterated H_3^+, *Astrophys. J.* **591**, pp. L41–L44.

Chapter 35

Dormand, J. R. and Woolfson, M. M. (1977) Interactions in the early solar system, *M. N. Roy. Astron. Soc.* **180**, pp. 243–279.

Woolfson, M. M. (2000) *The Origin and Evolution of the Solar System* (Institute of Physics, Bristol).

Chapter 36

Connell, A. J. and Woolfson, M. M. (1983) Evolution of the surface of Mars, *M. N. Roy. Astron. Soc.* **204**, pp. 1221–1240.

Guest, J., Butterworth, P., Murray, J. and O'Donnell, W. (1979) *Planetary Geology* (David and Charles, London).

Woolfson, M. M. (2000) *The Origin and Evolution of the Solar System* (Institute of Physics, Bristol).

Chapter 37

Benz, W., Slattery, W. L. and Cameron, A. G. W. (1986), The origin of the Moon and the single-impact hypothesis, *Icarus* **66**, pp. 515–535.
Woolfson, M. M. (1999) The Neptune-Triton-Pluto system, *M. N. Roy. Astron. Soc.* **304**, pp. 195–198.

Chapter 38

Hutchison, R. (1983) *A Search for our Beginning* (Oxford University Press, Oxford).

Chapter 39

Bailey, M. E., Clube, S. V. M. and Napier, W. M. (1990) *The Origin of Comets* (Butterworth–Heinemann, Oxford).

Chapter 40

Holden, P. and Woolfson, M. M. (1995) A theory of local formation of isotopic anomalies in meteorites, *Earth, Moon and Planets* **69**, pp. 201–236.

Index